BASICS OF ORGANIC CHEMISTRY

BASICS OF ORGANIC CHEMISTRY

DR. DURGESH KUMAR MAURYA

ANMOL PUBLICATIONS PVT. LTD.
NEW DELHI - 110 002 (INDIA)

ANMOL PUBLICATIONS PVT. LTD.

H.O.: 4374/4B, Ansari Road, Darya Ganj,
New Delhi-110 002 (India)
Ph.: 23278000, 23261597

B.O.: No. 1015, Ist Main Road, BSK IIIrd Stage
IIIrd Phase, IIIrd Block
Bangalore - 560 085 (India)
Visit us at: www.anmolpublications.com

Basics of Organic Chemistry

ISBN 978-81-261-3384-0

PRINTED IN INDIA

Printed at Mehra Offset Press, Delhi.

Contents

Preface

The past twenty years has seen an explosion of interest in free radicals as their pivotal role in both chemistry and biology has come to light. This introductory textbook aims to capture this excitement for advanced level undergraduates, with particular emphasis on the importance of radical reactions in organic synthesis.

The book provides a gentle, stepwise introduction to the subject, taking the student from the basic principles of radical reactions through to their applications in industry and their role in biological and environmental processes, allowing the relevance of the subject to be grasped more easily. Your chemistry course may include any of the five traditional branches of chemistry or a combination of two or more fields of chemistry:

- Inorganic chemistry - studies the structure & chemical reactions of substances composed of any of the known elements, except carbon-containing substances.
- Organic chemistry - studies of the compounds of carbon.
- Physical chemistry or theoretical chemistry - applies the application of theories and mathematical hERE methods to the solution of chemical problems.
- Analytical chemistry - deals with two areas: qualitative analysis (qual), "What is there?" and quantitative analysis (quant), "How much is there?"
- Biochemistry (or physiological chemistry) - studies the chemical structure of living material and the chemical reactions occurring in living cells. For example, general chemistry gives you an overview of each of the above five branches of chemistry.

Suitable for advanced level undergraduates and postgraduates in chemistry and biochemistry, the book will also be invaluable for research level scientists requiring an update in the area.

The development of university organic chemistry curricula and the trend towards modularisation of chemistry courses has driven the need for smaller, highly focused and accessible organic chemistry textbooks, which complement the very detailed "standard texts", to guide students through the key principles of the subject.

Author

Chapter 1

Introduction to Organic Chemistry

Organic chemistry is a specific discipline within the subject of chemistry. It is the scientific study of the structure, properties, composition, reactions, and preparation (by synthesis or by other means) of chemical compounds of carbon and hydrogen, which may contain any number of other elements, such as nitrogen, oxygen, halogens, and, more rarely, phosphorus or sulpher.

Organic chemicals get their diversity from the many different ways carbon can bond to other atoms.

Carbon (C) appears in the second row of the periodic table and has four bonding electrons in its valence shell. Similar to other non-metals, carbon needs eight electrons to satisfy its valence shell. Carbon therefore forms four bonds with other atoms (each bond consisting of one of carbon's electrons and one of the bonding atom's electrons). Every valence electron participates in bonding, thus a carbon atom's bonds will be distributed evenly over the atom's surface.

The simplest organic chemicals, called hydrocarbons, contain only carbon and hydrogen atoms; the simplest hydrocarbon (called methane) contains a single carbon atom bonded to four hydrogen atoms:

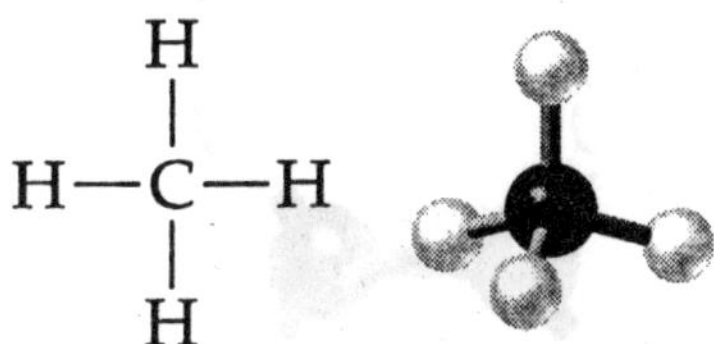

Fig. Methane

But carbon can bond to other carbon atoms in addition to hydrogen, as illustrated in the molecule ethane below:

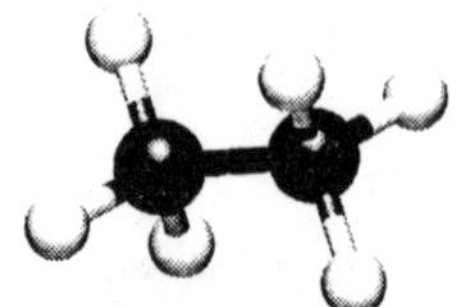
Fig. Ethane

In fact, the uniqueness of carbon comes from the fact that it can bond to itself in many different ways. Carbon atoms can form long chains:

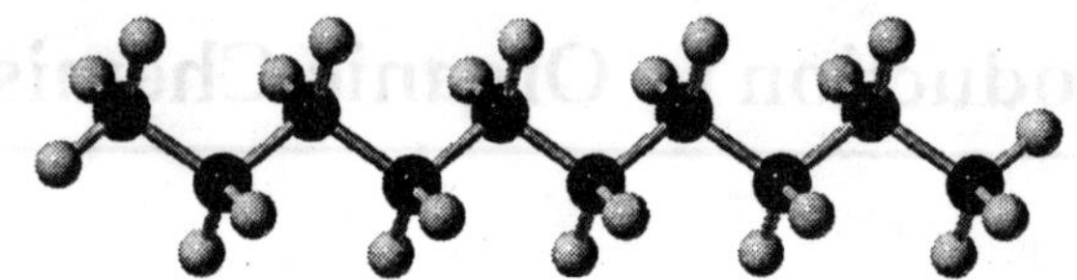
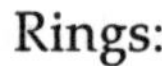
Fig. Decane

Branched Chains: Rings:

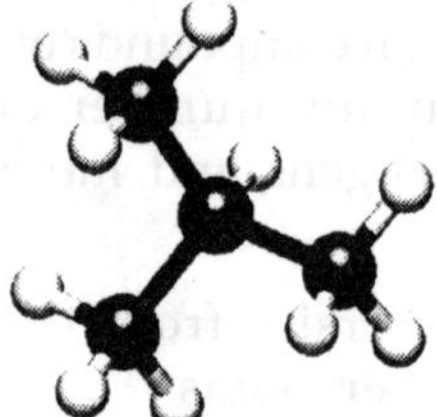
Fig. Iso butane

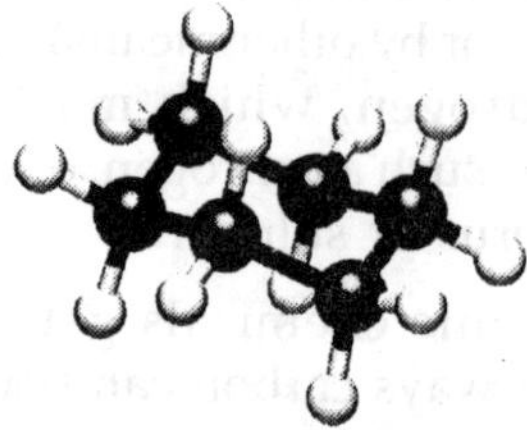
Fig. Cyclohexane

There appears to be almost no limit to the number of different structures that carbon can form. To add to the complexity of organic chemistry, neighboring carbon atoms can form double and triple bonds in addition to single carbon-carbon bonds:

Double Bounding: Triple Bounding:

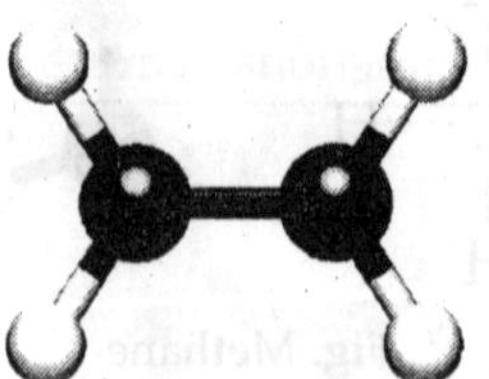
Fig. Ethene

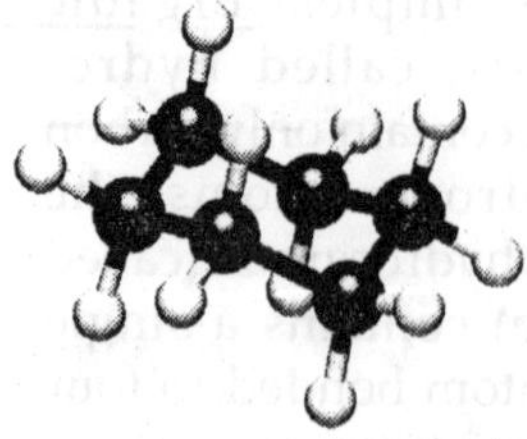
Fig. Ethyne

Hydrocarbons are divided into two groups - aliphatic and aromatic.

Aliphatic hydrocarbons can be simple like methane (1) (the main constituent of natural gas), which has only one carbon and four hydrogens, or very complex with branched or cyclic structures like cyclohexane.

Aliphatic compounds can be divided into categories depending on structural features.

They can be:

Alkanes, alkenes, alkynes, and cyclics.

Functional Groups

Functional groups are specific groups of atoms within molecules, that are responsible for the characteristic chemical reactions of those molecules. The same functional group will undergo the same or similar chemical reaction(s) regardless of the size of the molecule it is a part of. The term 'functional' group is linked to the concept of a homologous series. A homologous series is a group of molecules with the same general formula and the same functional group. They have similar physical and chemical properties (albeit with trends e.g. increasing boiling point with increasing carbon chain length). The terms higher/lower refer to a larger/smaller carbon chain.

Many important organic chemistry molecules contain oxygen or nitrogen. It's a good idea to memorize the names and structures of these functional groups.

–X, X= Cl, Br, I, F(*Alkyl Halide*)	Functional Group
–OH (*Alcohol*)	Functional Group
–O–(*Ether*)	Functional Group
—C—H ‖ C (*Aldehyde*)	Functional Group

Structure		
—C—OH with =O on C	(*Carboxylic Acid*)	Function Group
—C—O— with =O on C	(*Ester*)	Function Group
—N— (with bond up)	(*Amine*)	Function Group
—C—N— with =O on C	(*Amide*)	Function Group

IUPAC Nomenclature of Organic Compounds

The IUPAC nomenclature of organic chemistry is a systematic way of naming organic chemical compounds as recommended by the International Union of Pure and Applied Chemistry (IUPAC). Ideally, every organic compound should have a name from which an unambiguous structural formula can be drawn. For ordinary communication, to spare a tedious description, the official IUPAC naming recommendations are not always followed in practice except when it is necessary to give a concise definition to a compound, or when the IUPAC name is simpler (viz. ethanol against ethyl alcohol). Otherwise the common or trivial name may be used, often derived from the source of the compound.

(a) Nomenclature of Saturated Hydrocarbons

- Select the longest continuous chain of carbon atoms in the molecule. The compound is named as a derivative of this alkane
- Number the carbon atoms in the parent chain starting from the end which gives lowest possible sum for the numbers of the carbon atoms carrying the substituents
- That set of locants is preferred, which when

compared term by term with other set of locants, each in order of increasing magnitude, has the lowest term at the first point of difference. For example the set of locants (2,7,8) is preferred over the set of locants (3,4,9) since 2 comes before 3 even though the sum of locants in the former case is 17 while in the latter case, it is 16

```
1      2      3       4       5       6       7      8      9       10
Ch3 - Ch  - Ch2  - Ch2  - Ch2  - Ch2  - Ch  - Ch  - Ch2  - Ch3
       |                                      |      |
      Ch3                                    Ch3    Ch3
```

- The correct name is: 2,7,8 - Trimethyldecane and not 3,4,9 - Trimethyldecane

(b) Nomenclature of Compounds Containing Functional Group or Multiple Bonds

- Select the longest continuous chain containing the carbon atoms having the functional group or those involved in the multiple bonds
- The numbering of atoms in the parent chain is done in such a way that the carbon atom bearing the functional group or those carrying the multiple bond gets the lowest possible number
- While writing the name of alkene (double bond) or alkyne (triple bond), the primary suffix 'ane' of the corresponding alkane is replaced by 'enc' and 'yne' respectively. However, if the multiple bond occurs twice or thrice in the parent chain, the prefix di- or tri- is attached to the primary suffix ene or yne
- In naming the organic compounds containing one functional group a suffix known as secondary suffix is added to the primary suffix (giving number of carbon atoms in the chain) to indicate the nature of the functional group. A few important secondary suffixes are:

Functional group	Secondary suffix	Functional group	Secondary suffix

Alcohols (-OH)	-ol	Aldehydes (-CHO)	-al
Ketones (>C=O)	-one	Carboxylic acids (-COOH)	-oic acid
Amines ($-NH_2$)	-amine	Acid amides ($-CONH_2$)	-amide
Acid chlorides (-COCL)	-oyl chloride	Esters (-COOR)	-oate
Nitrites (-C≡N)	-nitrite	Thioalcohols (-SH)	-thiol

Nomenclature of Compounds having Polyfunctional Groups

When an organic compound contains two or more functional groups, one group is called the principal functional group while the others are called the secondary functional groups and are treated as substituents: The order of preference for principal group is: Carboxylic acid > acid anhydrides > esters > acid halides > amides > nitrites > aldehydes > ketone > alcohols > amines > double bond > triple bond.

When the functional groups act as substituents, they ar named as:

Functional group	Prefix	Functional group	Prefix
- COOH	Carboxy	-CHO	Formyl
-COOR	Alkoxy cabonyl or Carbalkoxy	>CO	Oxo or Keto
-COCL	Chloroformyl	-OH	Hydroxy
$-CONH_2$	Carbamoyl	-SH	Mecaplo
-CN	Cyano	$-NH_2$	Amino
-OR	Akoxy	=NH	Imino
-X	Halo	$-NO_2$	Nitro

Nomenclature of Simple Aromatic Compounds

(a) *Nuclear Substituted*: In these the functional group is directly attached to the benzene ring. Most of these compounds are better known by their common and historical names. In the IUPAC system, they are named as derivatives of benzene.

(b) *Side Chain Substituted*: In these the functional group is present in the side chain of the benzene ring. Both in the common and IUPAC systems, these are usually named as phenyl derivatives of the corresponding aliphatic compounds.

Chapter 2

Electronic Structure and Atomic Orbitals

A Simple View

The electronic structures of hydrogen and carbon can be drawn as:

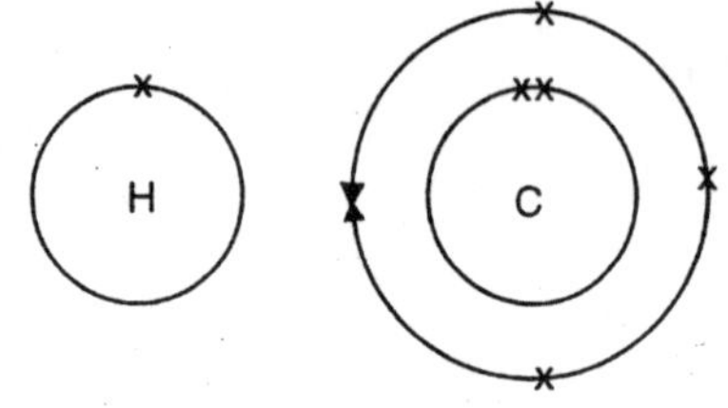

The circles show energy levels - representing increasing distances from the nucleus. You could straighten the circles out and draw the electronic structure as a simple energy diagram.

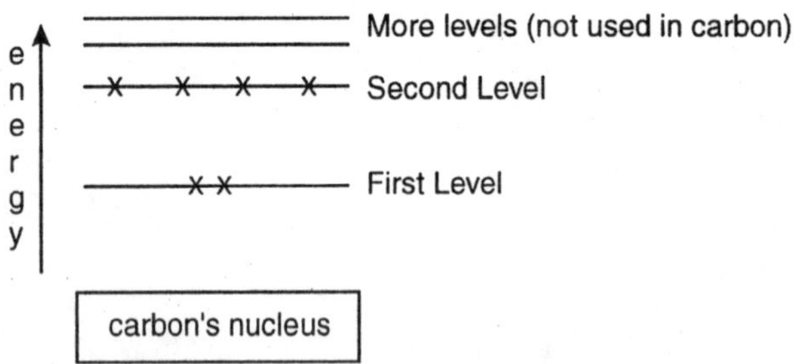

Atomic Orbitals

Orbits and orbitals sound similar, but they have quite different meanings. It is essential that you understand the difference between them.

The Impossibility of Drawing Orbits for Electrons

To plot a path for something you need to know exactly where the object is and be able to work out exactly where it's going to be an instant later. You can't do this for electrons.

The Heisenberg Uncertainty Principle says that you can't know with certainty both where an electron is and where it's going next. That makes it impossible to plot an orbit for an electron around a nucleus.

Hydrogen's Electron — The 1s Orbital

Suppose you had a single hydrogen atom and at a particular instant plotted the position of the one electron. Soon afterwards, you do the same thing, and find that it is in a new position. You have no idea how it got from the first place to the second.

You keep on doing this over and over again, and gradually build up a sort of 3D map of the places that the electron is likely to be found.

In the hydrogen case, the electron can be found anywhere within a spherical space surrounding the nucleus. The diagram shows a *cross-section* through this spherical space. 95% of the time (or any other percentage you choose), the electron will be found within a fairly easily defined region of space quite close to the nucleus. Such a region of space is called an orbital. You can think of an orbital as being the region of space in which the electron lives.

What is the electron doing in the orbital? We don't know, we can't know, and so we just ignore the problem! All you can say is that if an electron is in a particular orbital it will have a particular definable energy.

Each Orbital has a name.

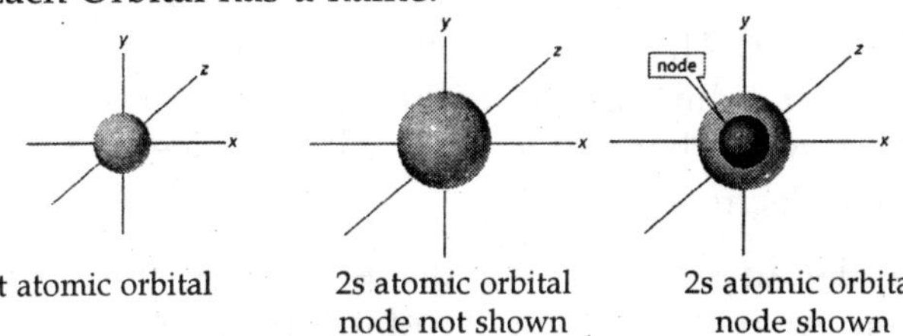

Fig. 1s and 2s orbitals

The orbital occupied by the hydrogen electron is called a 1s orbital. The "1" represents the fact that the orbital is in the energy level closest to the nucleus. The "s" tells you about the shape of the orbital. s orbitals are spherically symmetric around the nucleus - in each case, like a hollow ball made of rather chunky material with the nucleus at its centre.

The orbital above is a 2s orbital. This is similar to a 1s orbital except that the region where there is the greatest chance of finding the electron is further from the nucleus - this is an orbital at the second energy level.

If you look carefully, you will notice that there is another region of slightly higher electron density (where the dots are thicker) nearer the nucleus. ("Electron density" is another way of talking about how likely you are to find an electron at a particular place.)

2s (and 3s, 4s, etc) electrons spend some of their time closer to the nucleus than you might expect. The effect of this is to slightly reduce the energy of electrons in s orbitals. The nearer the nucleus the electrons get, the lower their energy.

3s, 4s (etc) orbitals get progressively further from the nucleus.

p orbitals

Not all electrons inhabit s orbitals (in fact, very few electrons live in s orbitals). At the first energy level, the only orbital available to electrons is the 1s orbital, but at the second level, as well as a 2s orbital, there are also orbitals called 2p orbitals.

A p orbital is rather like 2 identical balloons tied together at the nucleus. The diagram on the right is a cross-section through that 3-dimensional region of space. Once again, the orbital shows where there is a 95% chance of finding a particular electron.

Unlike an s orbital, a p orbital points in a particular direction. At any one energy level it is possible to have three absolutely equivalent p orbitals pointing mutually at right angles to each other. These are arbitrarily given the symbols p_x, p_y and p_z. This is simply for convenience - what you might think of as the x, y or z direction changes constantly as the atom tumbles in space.

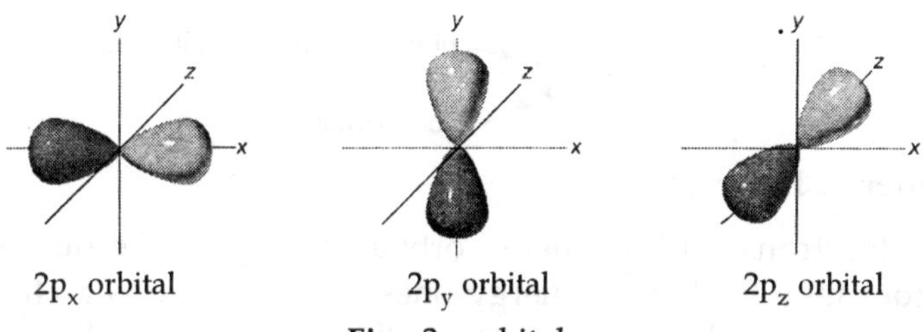

Fig. 2p orbitals

The p orbitals at the second energy level are called $2p_x$, $2p_y$ and $2p_z$. There are similar orbitals at subsequent levels - $3p_x$, $3p_y$, $3p_z$, $4p_x$, $4p_y$, $4p_z$ and so on.

All levels except for the first level have p orbitals. At the higher levels the lobes get more elongated, with the most likely place to find the electron more distant from the nucleus.

Fitting Electrons into Orbitals

Because for the moment we are only interested in the electronic structures of hydrogen and carbon, we don't need to concern ourselves with what happens beyond the second energy level.

Remember

At the first level there is only one orbital - the 1s orbital.

At the second level there are four orbitals - the 2s, $2p_x$, $2p_y$ and $2p_z$ orbitals.

Each orbital can hold either 1 or 2 electrons, but no more.

"Electrons-in-boxes"

Orbitals can be represented as boxes with the electrons in them shown as arrows. Often an up-arrow and a down-arrow are used to show that the electrons are in some way different.

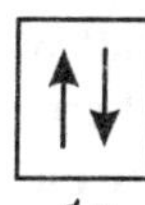

A 1s orbital holding 2 electrons would be drawn as shown on the right, but it can be written even more quickly as $1s^2$. This is read as "one s two" - not as "one s squared".

You mustn't confuse the two numbers in this notation:

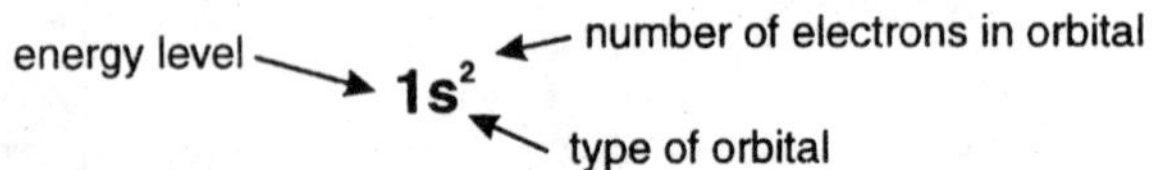

Order of Filling Orbitals

Electrons fill low energy orbitals (closer to the nucleus) before they fill higher energy ones. Where there is a choice between orbitals of equal energy, they fill the orbitals singly as far as possible.

The diagram (not to scale) summarises the energies of the various orbitals in the first and second levels.

Energy ordering of orbitals for multi-electron atoms

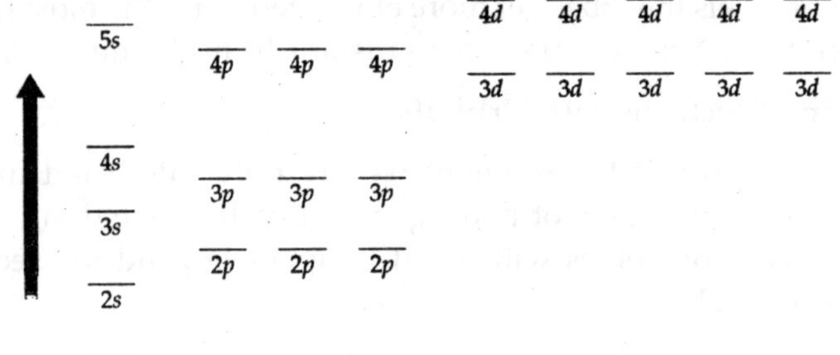

Fig. Energy of Orbitals

Notice that the 2s orbital has a slightly lower energy than the 2p orbitals. That means that the 2s orbital will fill with electrons before the 2p orbitals. All the 2p orbitals have exactly the same energy.

Electronic Structure of Hydrogen

Hydrogen only has one electron and that will go into the orbital with the lowest energy - the 1s orbital. Hydrogen has an electronic structure of $1s^1$. We have already described this orbital earlier.

H | 1 |

1s

Fig. Hydrogen atom

Electronic Structure of Carbon

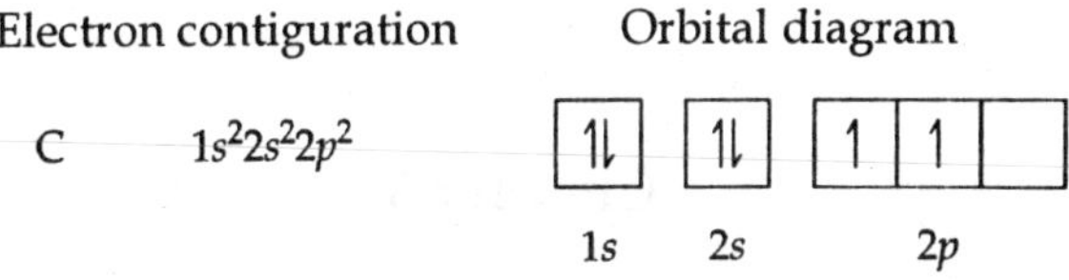

Fig. Carbon atom

Carbon has six electrons. Two of them will be found in the 1s orbital close to the nucleus. The next two will go into the 2s orbital. The remaining ones will be in two separate 2p orbitals. This is because the p orbitals all have the same energy and the electrons prefer to be on their own if that's the case.

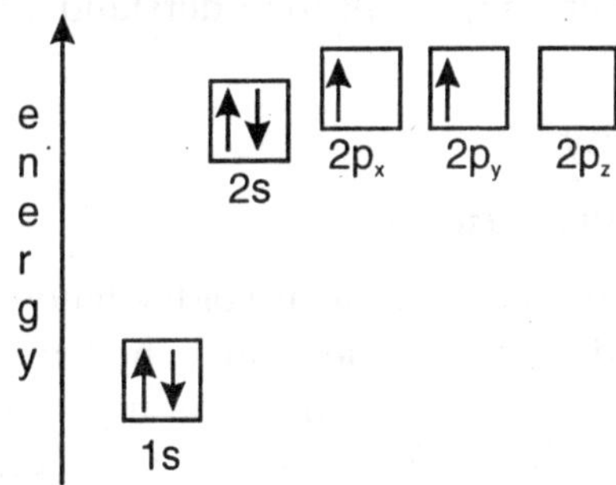

The electronic structure of carbon is normally written $1s^2 2s^2 2p_x^{\ 1} 2p_y^{\ 1}$.

Chapter 3

Bonding in Organic Molecules

Befor we discuss the concept of bonding in organic molecules it is very important to understand another concept, Electronegativity.

Electronegativity

What is Electronegativity?

Electronegativity is a measure of the tendency of an atom to attract a bonding pair of electrons. The Pauling scale is the most commonly used. Fluorine (the most electronegative element) is given a value of 4.0, and values range down to caesium and francium which are the least electronegative at 0.7.

What happens if two atoms of equal electronegativity bond together?

The most obvious example of this is the bond between two carbon atoms. Both atoms will attract the bonding pair to exactly the same extent. That means that on average the electron pair will be found half way between the two nuclei, and you could draw a picture of the bond like this:

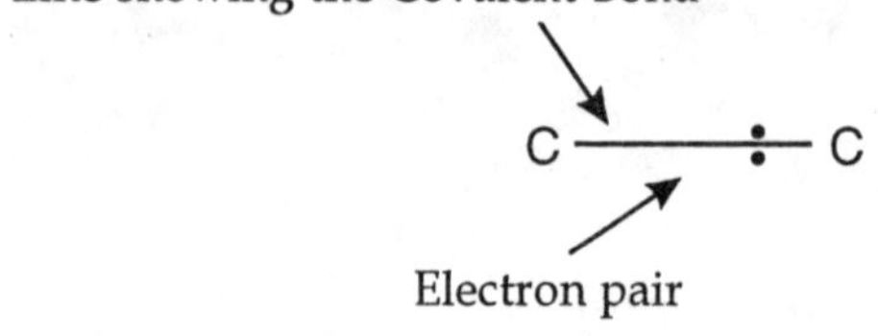

Line showing the Covalent Bond

C ——:— C

Electron pair

It is important to realize that this is an average picture. The electrons are actually in a sigma orbital, and are moving constantly within that orbital.

Carbon-fluorine Bond

Fluorine is much more electronegative than carbon. The actual values on the Pauling scale are

Carbon	2.5
Fluorine	4.0

That means that fluorine attracts the bonding pair much more strongly than carbon does. The bond - on average - will look like this:

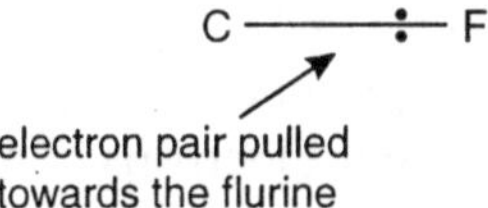

Why is Fluorine more Electronegative than Carbon?

A simple dots-and-crosses diagram of a C-F bond is perfectly adequate to explain it.

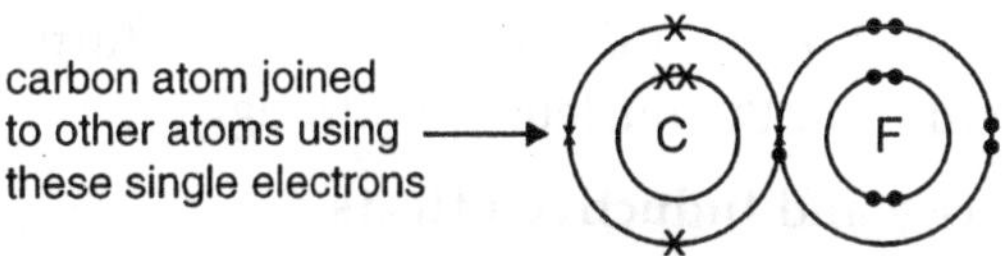

The bonding pair is in the second energy level of both carbon and fluorine, so in the absence of any other effect, the distance of the pair from both nuclei would be the same.

The electron pair is shielded from the full force of both nuclei by the 1s electrons - again there is nothing to pull it closer to one atom than the other.

BUT, the fluorine nucleus has 9 protons whereas the carbon nucleus has only 6.

Allowing for the shielding effect of the 1s electrons, the bonding pair feels a net pull of about 4+ from the carbon, but about 7+ from the fluorine. It is this extra nuclear charge which pulls the bonding pair (on average) closer to the fluorine than the carbon.

Carbon-chlorine Bond

The electronegativities are:

Carbon	2.5
Chlorine	3.0

The bonding pair of electrons will be dragged towards the chlorine but not as much as in the fluorine case. Chlorine isn't as electronegative as fluorine.

Why isn't Chlorine as Electronegative as Fluorine?

Chlorine is a bigger atom than fluorine.

Fluorine: $1s^2 2s^2 2p_x^{\,2} 2p_y^{\,2} 2p_z^{\,1}$

Chlorine: $1s^2 2s^2 2p_x^{\,2} 2p_y^{\,2} 2p_z^{\,2} 3s^2 3p_x^{\,2} 3p_y^{\,2} 3p_z^{\,1}$

In the chlorine case, the bonding pair will be shielded by all the 1-level and 2-level electrons. The 17 protons on the nucleus will be shielded by a total of 10 electrons, giving a net pull from the chlorine of about 7+.

That is the same as the pull from the fluorine, but with chlorine the bonding pair starts off further away from the nucleus because it is in the 3-level. Since it is further away, it feels the pull from the nucleus less strongly.

Bond Polarity and Inductive Effects

Polar Bonds

Think about the carbon-fluorine bond again. Because the bonding pair is pulled towards the fluorine end of the bond, that end is left rather more negative than it would otherwise be. The carbon end is left rather short of electrons and so becomes slightly positive.

$$\overset{\delta+}{C}\text{———}:\text{—}\overset{\delta-}{F}$$

The symbols δ+ and δ– mean "slightly positive" and "slightly negative". You read δ+ as "delta plus" or "delta positive".

We describe a bond having one end slightly positive and the other end slightly negative as being *polar*.

Inductive Effects

An atom like fluorine which can pull the bonding pair away from the atom it is attached to is said to have a negative inductive effect. Most atoms that you will come across have a negative inductive effect when they are attached to a carbon atom, because they are mostly more electronegative than carbon. You will come across some groups of atoms which have a slight positive inductive effect - they "push" electrons towards the carbon they are attached to, making it slightly negative.

Inductive effects are sometimes given symbols: *–I* (a negative inductive effect) and *+I* (a positive inductive effect).

Some Important Examples of Polar Bonds

Hydrogen bromide (and other hydrogen halides)

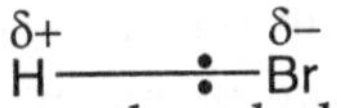

$$\overset{\delta+}{H}\text{———}:\overset{\delta-}{Br}$$

Bromine (and the other halogens) are all more electronegative than hydrogen, and so all the hydrogen halides have polar bonds with the hydrogen end slightly positive and the halogen end slightly negative.

The polarity of these molecules is important in their reactions with alkenes.

Carbon-bromine Bond in Halogenoalkanes

Bromine is more electronegative than carbon and so the bond is polarized in the way that we have already described with C-F and C-Cl.

$$\overset{\delta+}{C}\text{———}:\overset{\delta-}{Br}$$

The polarity of the carbon-halogen bonds is important in the reactions of the halogenoalkanes.

Carbon-oxygen Double Bond

The very electronegative oxygen atom pulls both bonding pairs towards itself - in the sigma bond and the pi bond. That leaves the oxygen fairly negative and the carbon fairly positive.

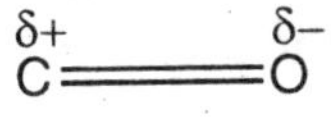

$$\overset{\delta+}{C}=\overset{\delta-}{O}$$

Chapter 4

Organic Reactions and their Mechanisms

Free radical Substitution

Substitution Reactions

These are reactions in which one atom in a molecule is replaced by another atom or group of atoms. Free radical substitution involves breaking a carbon-hydrogen bond in alkanes such as

Methane CH_4

Ethane CH_3CH_3

Propane $CH_3CH_2CH_3$

A new bond is then formed to something else

It also happens in alkyl groups like methyl, ethyl (and so on) wherever these appear in more complicated molecules.

Methyl CH_3

Ethyl CH_3CH_2

For example, ethanoic acid is CH_3COOH and contains a methyl group. The carbon-hydrogen bonds in the methyl group behave just like those in methane, and can be broken and replaced by something else in the same way.

A simple example of substitution is the reaction between methane and chlorine in the presence of UV light (or sunlight).

$$CH_4 + Cl_2 \longrightarrow CH_3Cl + HCl$$

Notice that one of the hydrogen atoms in the methane has been replaced by a chlorine atom. That's substitution.

Free Radical Reactions

Free radicals are atoms or groups of atoms which have a single unpaired electron. A free radical substitution reaction is one involving these radicals.

Free radicals are formed if a bond splits evenly - each atom getting one of the two electrons. The name given to this is homolytic fission.

To show that a species (either an atom or a group of atoms) is a free radical, the symbol is written with a dot attached to show the unpaired electron. For example:

A chlorine radical Cl

A methyl radical CH_3

Reaction between Methane and Chlorine

Here we are going to discuss the mechanism for the free radical substitution reaction between methane and chlorine.

If a mixture of methane and chlorine is exposed to a flame, it explodes - producing carbon and hydrogen chloride. This isn't a very useful reaction!

The reaction we are going to explore is a more gentle one between methane and chlorine in the presence of ultraviolet light - typically sunlight. This is a good example of a photochemical reaction - a reaction brought about by light.

$$CH_4 + Cl_2 \longrightarrow CH_3Cl + HCl$$

The organic product is chloromethane.

One of the hydrogen atoms in the methane has been replaced by a chlorine atom, so this is a substitution reaction. However, the reaction doesn't stop there, and all the hydrogens in the methane can in turn be replaced by chlorine atoms.

Mechanism

The mechanism involves a chain reaction. During a chain reaction, for every reactive species you start off with, a new one is generated at the end - and this keeps the process going.

The over-all process is known as free radical substitution, or as a free radical chain reaction.

Chain Initiation

The chain is initiated (started) by UV light breaking a chlorine molecule into free radicals.

$$Cl_2 \longrightarrow 2Cl\bullet$$

Chain Propagation Reactions

These are the reactions which keep the chain going.

Chain propagation reactions

These are the reactions which keep the chain going.

$$CH_4 + Cl\bullet \longrightarrow CH_3\bullet + HCl$$

$$CH_3\bullet + Cl_2 \longrightarrow CH_3Cl + Cl\bullet$$

Chain Termination Reactions

These are reactions which remove free radicals from the system without replacing them by new ones.

$$2Cl\bullet \longrightarrow Cl_2$$

$$CH_3\bullet + Cl\bullet \longrightarrow CH_3Cl$$

$$CH_3\bullet + CH_3\bullet \longrightarrow CH_3CH_3$$

Reaction between Methane and Bromine

This reaction between methane and bromine happens in the presence of ultraviolet light - typically sunlight. This is a good example of a photochemical reaction - a reaction brought about by light.

$$CH_4 + Br_2 \longrightarrow CH_3Br + HBr$$

The organic product is bromomethane.

One of the hydrogen atoms in the methane has been replaced by a bromine atom, so this is a substitution reaction. However, the reaction doesn't stop there, and all the hydrogens in the methane can in turn be replaced by bromine atoms.

Mechanism

The mechanism involves a chain reaction. During a chain reaction, for every reactive species you start off with, a new one is generated at the end - and this keeps the process going.

The over-all process is known as free radical substitution, or as a free radical chain reaction.

Chain initiation

The chain is initiated (started) by UV light breaking a bromine molecule into free radicals.

$$Br_2 \longrightarrow 2Br\bullet$$

Chain Propagation Reactions

These are the reactions which keep the chain going.

$$CH_4 + Br\bullet \longrightarrow CH_3\bullet + HBr$$

$$CH_3\bullet + Br_2 \longrightarrow CH_3Br + Br\bullet$$

Chain Termination Reactions

These are reactions which remove free radicals from the system without replacing them by new ones.

$$2Br\bullet \longrightarrow Br_2$$

$$CH_3\bullet + Br\bullet \longrightarrow CH_3Br$$

$$CH_3\bullet + CH_3\bullet \longrightarrow CH_3CH_3$$

Electrophilic Addition

Background

Electrophiles

An electrophile is something which is attracted to electron-rich regions in other molecules or ions. Because it is attracted to a negative region, an electrophile must be something which carries either a full positive charge, or has a slight positive charge on it somewhere.

Ethene and the other alkenes are attacked by electrophiles. The electrophile is normally the slightly positive (δ+) end of a molecule like hydrogen bromide, HBr.

Electrophiles are strongly attracted to the exposed electrons in the pi bond and reactions happen because of that initial attraction - as you will see shortly.

$$\overset{\delta+}{H}\text{---}:\text{---}\overset{\delta-}{Br}$$

One might wonder why fully positive ions like sodium, Na^+, don't react with ethene. Although these ions may well be attracted to the pi bond, there is no possibility of the process going any further to form bonds between sodium and carbon, because sodium forms ionic bonds, whereas carbon normally forms covalent ones.

Addition Reactions

In a sense, the pi bond is an unnecessary bond. The structure would hold together perfectly well with a single bond rather than a double bond. The pi bond often breaks and the electrons in it are used to join other atoms (or groups of atoms) onto the ethene molecule. In other words, ethene undergoes addition reactions.

For example, using a general molecule X–Y . . .

$$H_2C{=}CH_2 + X{-}Y \longrightarrow H{-}\underset{X}{\overset{H}{C}}{-}\underset{Y}{\overset{H}{C}}{-}H$$

Electrophilic Addition Reactions

An addition reaction is a reaction in which two molecules join together to make a bigger one. Nothing is lost in the process. All the atoms in the original molecules are found in the bigger one.

An electrophilic addition reaction is an addition reaction which happens because what we think of as the "important" molecule is attacked by an electrophile. The "important" molecule has a region of high electron density which is attacked by something carrying some degree of positive charge.

Electrophilic addition reaction occurs in many of the reactions of compounds containing carbon-carbon double bonds - the alkenes.

The Structure of Ethene

Lets start by looking at ethene, because it is the simplest molecule containing a carbon-carbon double bond. What is true of C=C in ethene will be equally true of C=C in more complicated alkenes.

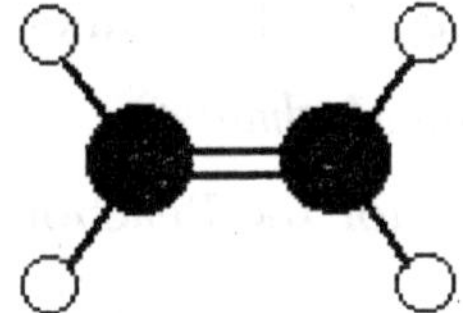

Ethene, C_2H_4, is often modeled as shown above. The double bond between the carbon atoms is, of course, two pairs of shared electrons. What the diagram doesn't show is that the two pairs aren't the same as each other.

One of the pairs of electrons is held on the line between the two carbon nuclei as one would expect, but the other is held in a molecular orbital above and below the plane of the molecule. A molecular orbital is a region of space within the molecule where there is a high probability of finding a particular pair of electrons.

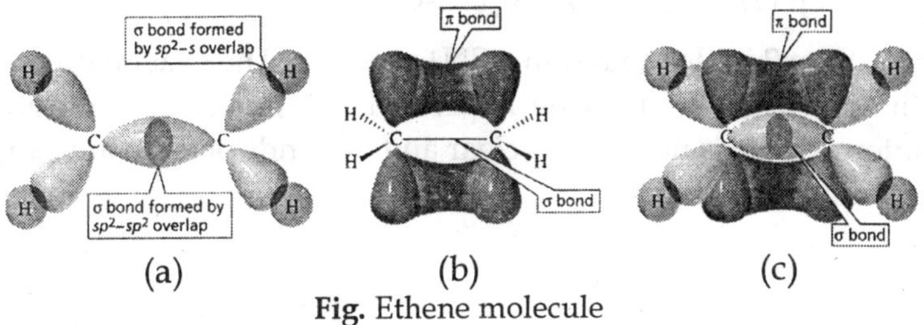

Fig. Ethene molecule

In this diagram, the line between the two carbon atoms represents a normal bond - the pair of shared electrons lies in a molecular orbital on the line between the two nuclei where you would expect them to be. This sort of bond is called a sigma bond.

The other pair of electrons is found somewhere in the shaded part above and below the plane of the molecule. This bond is called a pi bond. The electrons in the pi bond are free to move around anywhere in this shaded region and can move freely from one half to the other.

The pi electrons are not as fully under the control of the carbon nuclei as the electrons in the sigma bond and, because they lie exposed above and below the rest of the molecule, they are relatively open to attack by other things.

Electrophilic Addition Mechanism

The Mechanism for the Reaction between Ethene and a Molecule X-Y

The driving force for this reaction is the formation of an electrophile Y^+ that forms a covalent bond with an electron-rich unsaturated C=C bond.

The positive charge on Y is transferred to the carbon-carbon bond.

Step (1) Y^+ + -C=C- $\rightarrow$ Y-C-C^+-

In step 2 of an Electrophilic addition, the positively charged intermediate combines with (Z) that is electron-rich to form the second covalent bond.

Step (2) YC-C^+- + Z $\rightarrow$ Y-C-C-Z

Step 2 is also found in a SN1 reaction. The exact nature of the electrophile and the nature of the positively charged intermediate is not always clear and depends on reactants and reaction conditions.

An ion in which the positive charge is carried on a carbon atom is called a carbocation or a carbonium ion (an older term).

Reaction between Symmetrical Alkenes and Sulphuric Acid

The electrophilic addition reaction between ethene and sulphuric acid

Alkenes react with concentrated sulphuric acid in the cold to produce alkyl hydrogensulphates. Ethene reacts to give ethyl hydrogensulphate.

$$CH_2{=}CH_2 + H_2SO_4 \longrightarrow CH_3CH_2OSO_2OH$$

The structure of the product molecule is sometimes written as $CH_3CH_2HSO_4$, but the version in the equation is better because it shows how all the atoms are linked up.

The Mechanism for the Reaction between Ethene and Sulphuric Acid

Sulphuric Acid as an Electrophile

The hydrogen atoms are attached to very electronegative oxygen atoms which means that the hydrogens will have a slight positive charge while the oxygens will be slightly negative. In the mechanism, we just focus on one of the hydrogen to oxygen bonds, because the other one is too far from the carbon-carbon double bond to be involved in any way.

The Mechanism

CH2=CH2 + H(δ+)–O(δ-)–SO2–OH ⟶ CH2–C⁺H2 (with H) + ⁻O–SO2–OH ⟶ CH2(H)–CH2–O–SO2–OH

Electrophilic Addition Reaction between Cyclohexene and Sulphuric Acid

Here we are going straight for the mechanism without producing an initial equation.

The Mechanism for the Reaction between Cyclohexene and Sulphuric Acid

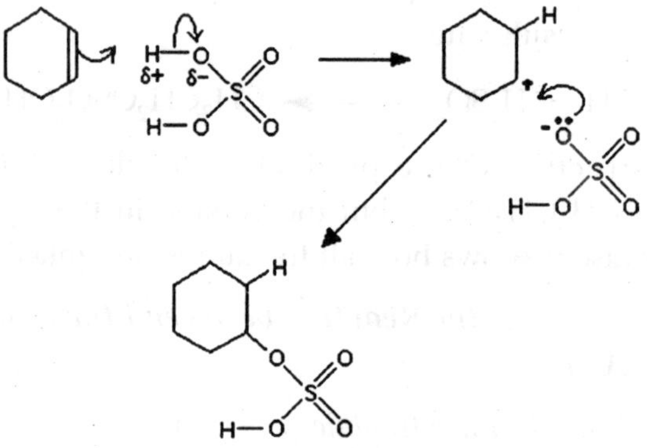

Having worked out the structure of the product, one could now write a simple equation for the reaction if one wants to.

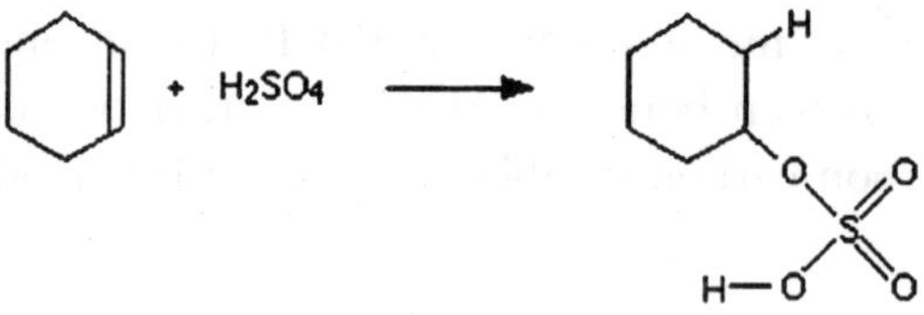

Reaction between Symmetrical Alkenes and the Hydrogen Halides

Hydrogen halides include hydrogen chloride and hydrogen bromide.

Electrophilic Addition Reactions Involving Hydrogen Bromide

Alkenes react with hydrogen bromide in the cold. The double bond breaks and a hydrogen atom ends up attached to one of the carbons and a bromine atom to the other.

In the case of ethene, bromoethane is formed.

$$CH_2 = CH_2 + HBr \longrightarrow CH_3CH_2Br$$

With cyclohexene we get bromocyclohexane.

CH2, CH2, CH, CH, CH2, CH2 (ring) + HBr ⟶ CH2, CH2, CH2, CHBr, CH2, CH2 (ring)

The structures of the cyclohexene and the bromocyclohexane are often simplified:

(cyclohexene) + HBr ⟶ (bromocyclohexane: H, Br)

Mechanism

The reactions are examples of electrophilic addition.

With ethene and HBr:

$$CH_2{=}CH_2 + H^{\delta+}{-}Br^{\delta-} \longrightarrow CH_2{-}\overset{+}{C}H_2 \text{ (H on first C)} + :Br^- \longrightarrow CH_2(H){-}CH_2(Br)$$

and with cyclohexene:

(cyclohexene) + $H^{\delta+}{-}Br^{\delta-}$ ⟶ (carbocation, H) + $:Br^-$ ⟶ (bromocyclohexane: H, Br)

Electrophilic Addition Reactions Involving the Other Hydrogen Halides

Hydrogen chloride and the other hydrogen halides add on in exactly the same way. For example, hydrogen chloride adds to ethene to make chloroethane:

$$CH_2{=}CH_2 + HCl \longrightarrow CH_3CH_2Cl$$

The only difference is in how fast the reactions happen with the different hydrogen halides. The rate of reaction increases as you go from HF to HCl to HBr to HI.

HF	Slowest reaction
HCl	
HBr	
HI	Fastest reaction

The reason for this is that as the halogen atoms get bigger, the strength of the hydrogen-halogen bond falls. Bond strengths (measured in kilojoules per mole) are:

H-F	568
H-Cl	432
H-Br	366
H-I	298

As you have seen in the HBr case, in the first step of the mechanism the hydrogen-halogen bond gets broken. If the bond is weaker, it will break more readily and so the reaction is more likely to happen.

Mechanism

The reactions are still examples of *electrophilic addition.*

With ethene and HCl, for example:

$$CH_2{=}CH_2 + H^{\delta+}{-}Cl^{\delta-} \longrightarrow CH_2(H){-}\overset{+}{C}H_2 + :Cl^- \longrightarrow CH_2(H){-}CH_2(Cl)$$

This is exactly the same as the mechanism for the reaction between ethene and HBr, except that we've replaced Br by Cl.

All the other mechanisms for symmetrical alkenes and the hydrogen halides would be done in the same way.

Reaction between Symmetrical Alkenes and Bromine

Electrophilic Addition of Bromine to Ethene

Alkenes react in the cold with pure liquid bromine, or with a solution of bromine in an organic solvent like tetrachloromethane. The double bond breaks, and a bromine atom becomes attached to each carbon. The bromine loses its original red-brown colour to give a colourless liquid. In the case of the reaction with ethene, 1,2-dibromoethane is formed.

$$CH_2{=}CH_2 + Br_2 \longrightarrow \underset{Br}{CH_2}{-}\underset{Br}{CH_2}$$

This decolourisation of bromine is often used as a test for a carbon-carbon double bond. If an aqueous solution of bromine is used ("bromine water"), you get a mixture of products.

The other halogens, apart from fluorine, behave similarly. (Fluorine reacts explosively with all hydrocarbons - including alkenes - to give carbon and hydrogen fluoride.)

The Mechanism for the Reaction between Ethene and Bromine

The reaction proceeds by *electrophilic addition.*

Bromine as an Electrophile

The bromine is a very "polarisable" molecule and the approaching pi bond in the ethene induces a dipole in the bromine molecule.

The Simplified Version of the Mechanism

$CH_2{=}CH_2$ + $Br^{\delta+}{-}Br^{\delta-}$ (induced dipole) $\longrightarrow$ $\underset{Br}{CH_2}{-}\overset{+}{C}H_2$ + Br^- $\longrightarrow$ $\underset{Br}{CH_2}{-}\underset{Br}{CH_2}$

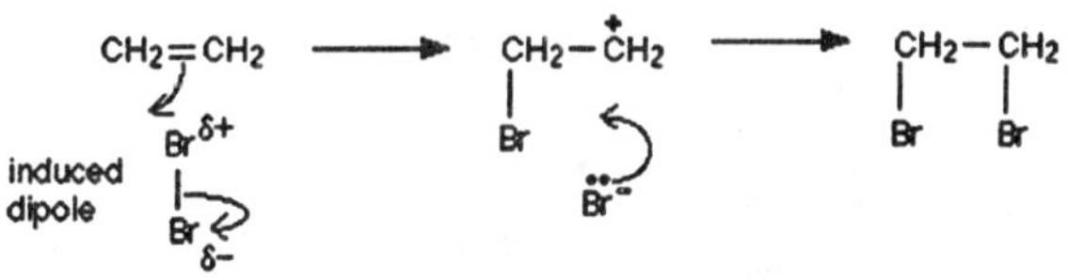

The more Accurate Version of the Mechanism

In the first stage of the reaction, one of the bromine atoms becomes attached to both carbon atoms, with the positive charge being found on the bromine atom. A bromonium ion is formed.

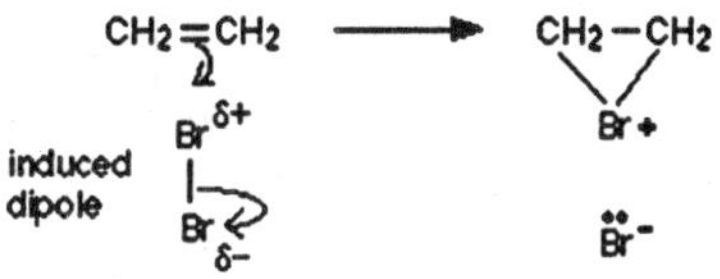

$CH_2{=}CH_2$ + $Br^{\delta+}{-}Br^{\delta-}$ (induced dipole) $\longrightarrow$ $CH_2{-}CH_2$ bridged by Br^+, + Br^-

The bromonium ion is then attacked from the back by a bromide ion formed in a nearby reaction.

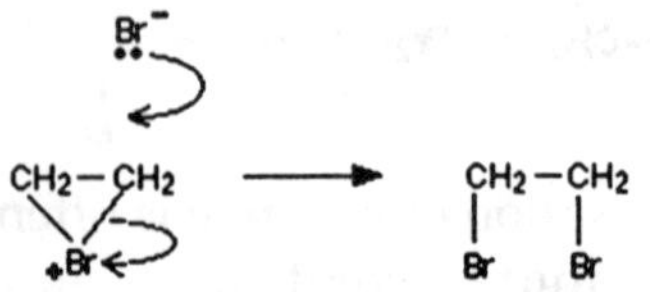

Electrophilic Addition of bromine to cyclohexene

Cyclohexene reacts with bromine in the same way and under the same conditions as any other alkene. 1,2-dibromocyclohexane is formed.

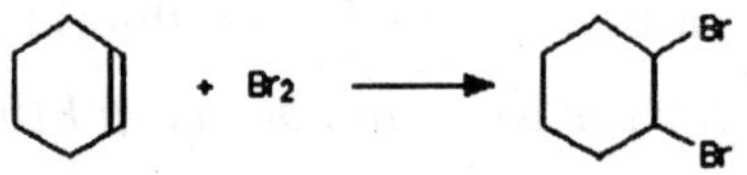

The Mechanism for the Reaction between Cyclohexene and Bromine

The reaction proceeds by *electrophilic addition.*

Bromine as an Electrophile

Again, the bromine is polarised by the approaching pi bond in the cyclohexene. Don't forget to write the words "induced dipole" next to the bromine molecule.

The Simplified Version of the Mechanism

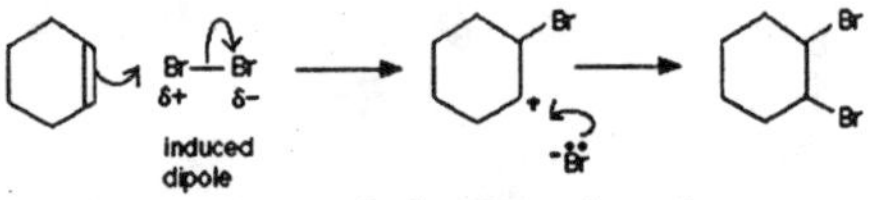

The Alternative Version of the Mechanism

In the first stage of the reaction, one of the bromine atoms becomes attached to both carbon atoms, with the positive charge being found on the bromine atom. A bromonium ion is formed.

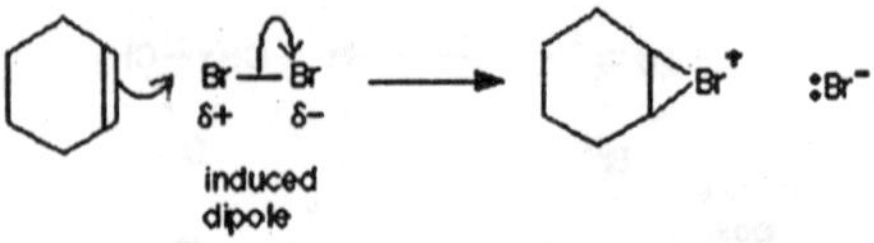

The bromonium ion is then attacked from the back by a bromide ion formed in a nearby reaction.

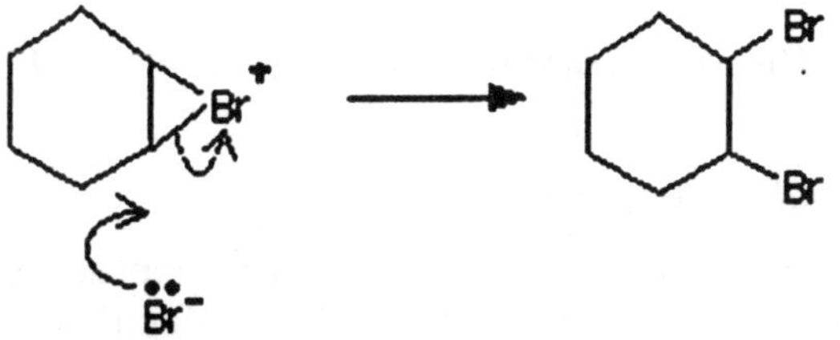

Electrophilic Substitution

Electrophilic substitution happens in many of the reactions of compounds containing benzene rings - the arenes. For simplicity, we'll only look for now at benzene itself.

This is what you need to understand for the purposes of the electrophilic substitution mechanisms:

- Benzene, C_6H_6, is a planar molecule containing a ring of six carbon atoms each with a hydrogen atom attached.
- There are delocalized electrons above and below the plane of the ring.

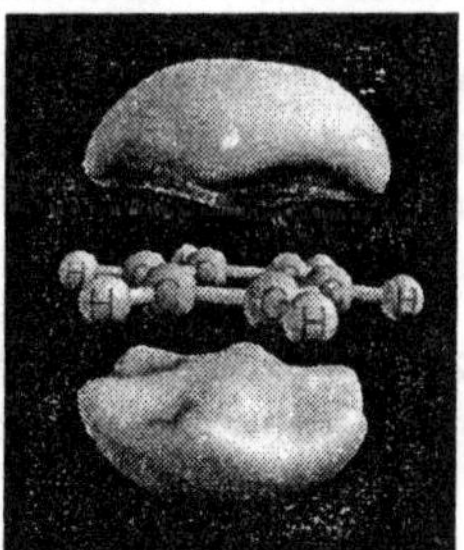

Fig. Benzene molecule

- The presence of the delocalized electrons makes benzene particularly stable.
- Benzene resists addition reactions because that would involve breaking the delocalization and losing that stability.

Benzene is represented by this symbol, where the circle represents the delocalized electrons, and each corner of the hexagon has a carbon atom with a hydrogen attached.

Electrophilic Substitution Reactions Involving Positive Ions

Benzene and Electrophiles

Because of the delocalized electrons exposed above and below the plane of the rest of the molecule, benzene is obviously going to be highly attractive to electrophiles - species which seek after electron rich areas in other molecules.

The electrophile will either be a positive ion, or the slightly positive end of a polar molecule.

The delocalized electrons above and below the plane of the benzene molecule are open to attack in the same way as those above and below the plane of an ethene molecule. However, the end result will be different.

If benzene underwent addition reactions in the same way as ethene, it would need to use some of the delocalized electrons to form bonds with the new atoms or groups. This would break the delocalization - and this costs energy.

Instead, it can maintain the delocalization if it replaces a hydrogen atom by something else - a substitution reaction. The hydrogen atoms aren't involved in any way with the delocalized electrons.

In most of benzene's reactions, the electrophile is a positive ion, and these reactions all follow a general pattern.

General Mechanism

The First Stage

Suppose the electrophile is a positive ion X^+.

Two of the electrons in the delocalised system are attracted towards the X^+ and form a bond with it. This has the effect of breaking the delocalization, although not completely.

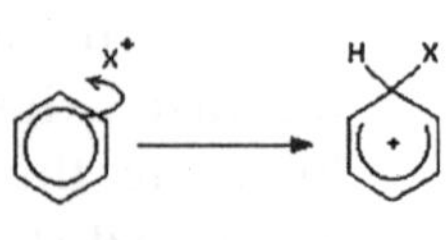

The ion formed in this step isn't the final product. It immediately goes on to react with something else. It is just an intermediate.

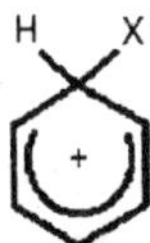

There is still delocalization in the intermediate formed, but it only covers part of the ion. When you write one of these mechanisms, draw the partial delocalization to take in all the carbon atoms apart from the one that the X has become attached to.

The intermediate ion carries a positive charge because you are joining together a neutral molecule and a positive ion. This positive charge is spread over the delocalised part of the ring. Simply draw the "+" in the middle of the ring.

The hydrogen at the top isn't new - it's the hydrogen that was already attached to that carbon. We need to show that it is there for the next stage.

The Second Stage

Here we've introduced a new ion, Y^-. It is simply the negative ion that was originally associated with X^+.

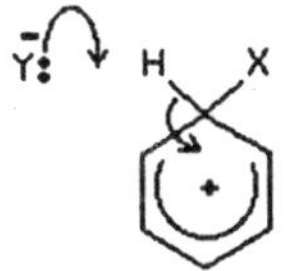

A lone pair of electrons on Y^- forms a bond with the hydrogen atom at the top of the ring. That means that the pair of electrons joining the hydrogen onto the ring aren't needed any more. These then move down to plug the gap in the delocalised electrons, so restoring the delocalised ring of electrons which originally gave the benzene its special stability.

The Energetics of the Reaction

The complete delocalization is temporarily broken as X replaces H on the ring, and this costs energy. However, that energy is recovered when the delocalization is re-established. This initial input of energy is simply the activation energy for the reaction. In this case, it is going to be high (something around 150 kJ mol^{-1}), and this means that benzene's reactions tend to be slow.

Electrophilic Substitution Reactions not Involving Positive Ions

Halogenation and Sulphonation

In these reactions, the electrophiles are polar molecules rather than fully positive ions.

Nucleophilic Substitution

Nucleophiles

A nucleophile is a species (an ion or a molecule) which is strongly attracted to a region of positive charge in something else.

Nucleophiles are either fully negative ions, or else have a strongly δ- charge somewhere on a molecule. Common nucleophiles are hydroxide ions, cyanide ions, water and ammonia.

$$^{-}:\ddot{O}-H \qquad ^{-}:C\equiv N \qquad :\overset{\delta-}{\ddot{O}}-\overset{\delta+}{H} \;(\text{with } {}^{\delta+}H) \qquad \overset{\delta+}{H}-\overset{\delta-}{\ddot{N}}-\overset{\delta+}{H} \;(\text{with } {}^{\delta+}H)$$

Notice that each of these contains at least one lone pair of electrons, either on an atom carrying a full negative charge, or on a very electronegative atom carrying a substantial - charge.

The nucleophiles that we shall be looking at all depend on lone pairs on either an oxygen atom or a nitrogen atom.

Nucleophiles based on Oxygen Atoms

We shall be talking about water and alcohols, taking ethanol as a typical alcohol.

$$\begin{array}{c} H-\ddot{O}: \\ | \\ H \end{array} \qquad \begin{array}{r} CH_3CH_2-\ddot{O}: \\ | \;\; \\ H \;\; \end{array}$$

Notice how similar these two molecules are around the oxygen atom. That's what turns out to be important.

Nucleophiles based on Nitrogen Atoms

We shall be considering ammonia and primary amines, taking ethylamine as a typical primary amine. A primary amine contains the -NH_2 group attached to either an alkyl group (as it is here) or a benzene ring. As far as these reactions are concerned, the nature of any hydrocarbon attached to the nitrogen makes no difference. The nitrogen atom is the important bit.

$$\begin{array}{c} H-\ddot{N}-H \\ | \\ H \end{array} \qquad \begin{array}{r} CH_3CH_2-\ddot{N}-H \\ | \;\;\;\;\; \\ H \;\;\;\;\; \end{array}$$

Again, notice how similar these two molecules are around the nitrogen atom - and also how similar they are to the previous ones containing oxygen. Both types of molecule have an active lone pair of electrons attached to one of the most electronegative elements. All of these molecules also have at least one hydrogen atom attached to the oxygen or nitrogen.

Bonding in the Halogenoalkanes

Halogenoalkanes (also known as haloalkanes or alkyl halides) are compounds containing a halogen atom (fluorine, chlorine, bromine or iodine) joined to one or more carbon atoms in a chain.

The interesting thing about these compounds is the carbon-halogen bond, and all the nucleophilic substitution reactions of the halogenoalkanes involve breaking that bond.

Polarity of the Carbon-halogen Bonds

With the exception of iodine, all of the halogens are more electronegative than carbon.

Electronegativity Values (Pauling Scale)

C	2.5	F	4.0
		Cl	3.0
		Br	2.8
		I	2.5

That means that the electron pair in the carbon-halogen bond will be dragged towards the halogen end, leaving the halogen slightly negative ($\delta-^{-}$) and the carbon slightly positive ($\delta+$) - except in the carbon-iodine case.

Although the carbon-iodine bond doesn't have a permanent dipole, the bond is very easily polarised by anything approaching it. Imagine a negative ion approaching the bond from the far side of the carbon atom:

$$Nu^{-} \quad -\overset{\delta+}{C}-\overset{\delta-}{LG} \longrightarrow -C-Nu \quad LG^{-}$$

The fairly small polarity of the carbon-bromine bond will be increased by the same effect.

The fairly small polarity of the carbon-bromine bond will be increased by the same effect.

Strengths of the Carbon-halogen Bonds

Look at the strengths of various bonds (all values in kJ mol^{-1}).

C-H	413	C-F	467
		C-Cl	346
		C-Br	290
		C-I	228

In all of these nucleophilic substitution reactions, the carbon-halogen bond has to be broken at some point during the reaction. The harder it is to break, the slower the reaction will be.

The carbon-fluorine bond is very strong (stronger than C-H) and isn't easily broken. It doesn't matter that the carbon-fluorine bond has the greatest polarity - the strength of the bond is much more important in determining its reactivity. You might therefore expect fluoroalkanes to be very unreactive - and they are!

In the other halogenoalkanes, the bonds get weaker as you go from chlorine to bromine to iodine.

That means that chloroalkanes react most slowly, bromoalkanes react faster, and iodoalkanes react faster still.

Rates of Reaction: RCl < RBr < RI

Where "<" is read as "is less than" - or, in this instance, "is slower than", and **R** represents any alkyl group.

Nucleophilic Substitution in Primary Halogenoalkanes

The Nucleophilic Substitution Reaction - an S_N2 Reaction

We'll talk this mechanism through using an ion as a nucleophile, because it's slightly easier. The water and ammonia mechanisms involve an extra step which you can read about on the pages describing those particular mechanisms.

We'll take bromoethane as a typical primary halogenoalkane. The bromoethane has a polar bond between the carbon and the bromine.

We'll look at its reaction with a general purpose nucleophilic ion which we'll call Nu^-. This will have at least one lone pair of electrons. Nu^- could, for example, be OH^- or CN^-.

The lone pair on the Nu^- ion will be strongly attracted to the δ+ carbon, and will move towards it, beginning to make a co-ordinate (dative covalent) bond. In the process the electrons in the C-Br bond will be pushed even closer towards the bromine, making it increasingly negative.

The Nu^- ion approaches the δ+ carbon from the side away from the bromine atom. The large bromine atom hinders attack from its side and, being -, would repel the incoming Nu^- anyway. This attack from the back is important if you need to understand why tertiary halogenoalkanes have a different mechanism. We'll discuss this later on this page.

There is obviously a point in which the Nu^- is half attached to the carbon, and the C-Br bond is half way to being broken. This is called a transition state. It isn't an intermediate. You can't isolate it - even for a very short time. It's just the mid-point of a smooth attack by one group and the departure of another.

Mechanism

The simplest way is:

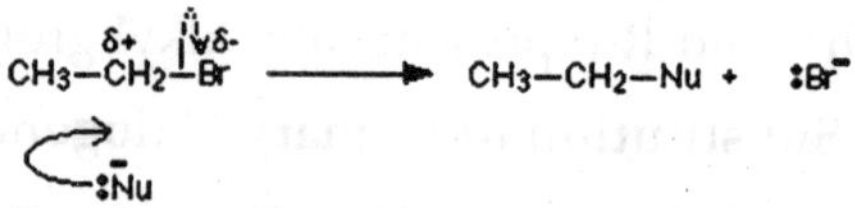

$$CH_3-\overset{\delta+}{C}H_2-\overset{\delta-}{Br} \longrightarrow CH_3-CH_2-Nu + :Br^-$$

$$:\bar{N}u$$

Technically, this is known as an S_N2 reaction. S stands for substitution, N for nucleophilic, and the 2 is because the initial stage of the reaction involves two species - the bromoethane and the Nu^- ion. If your syllabus doesn't refer to S_N2 reactions by name, you can just call it *nucleophilic substitution.*

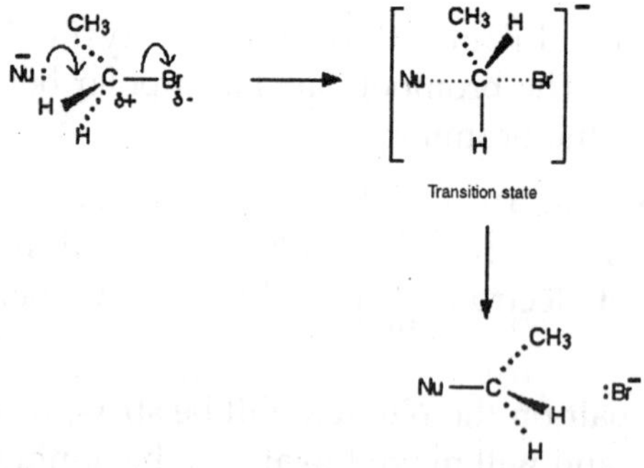

$$\bar{N}u: \; H\overset{CH_3}{\underset{H}{C}}{}^{\delta+}-Br^{\delta-} \longrightarrow \left[Nu\cdots\overset{CH_3\;H}{\underset{H}{C}}\cdots Br \right]^-$$

Transition state

$$\downarrow$$

$$Nu-\overset{CH_3}{\underset{H}{C}}-H \qquad :Br^-$$

Be very careful when you draw the transition state to make a clear difference between the dotted lines showing the half-made and half-broken bonds, and those showing the bonds going back into the paper.

Notice that the molecule has been inverted during the reaction - rather like an umbrella being blown inside-out.

Nucleophilic Substitution in Tertiary Halogenoalkanes

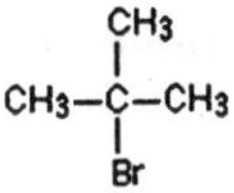

$$CH_3-\overset{\displaystyle CH_3}{\underset{\displaystyle Br}{\overset{|}{\underset{|}{C}}}}-CH_3$$

Remember that a tertiary halogenoalkane has three alkyl groups attached to the carbon with the halogen on it. These alkyl groups can be the same or different, but in this section, we shall just consider a simple one, $(CH_3)_3CBr$ - 2-bromo-2-methylpropane.

The Nucleophilic Substitution Reaction - an S_N1 Reaction

Once again, we'll talk this mechanism through using an ion as a nucleophile, because it's slightly easier, and again we'll look at the reaction of a general purpose nucleophilic ion which we'll call Nu^-. This will have at least one lone pair of electrons.

Why we need a Different Mechanism?

You will remember that when a nucleophile attacks a primary halogenoalkane, it approaches the $\delta+$ carbon atom from the side away from the halogen atom.

With a tertiary halogenoalkane, this is impossible. The back of the molecule is completely cluttered with CH_3 groups.

Since any other approach is prevented by the bromine atom, the reaction has to go by an alternative mechanism.

The Alternative Mechanism

The reaction happens in two stages. In the first, a small proportion of the halogenoalkane ionizes to give a carbocation and a bromide ion.

$$CH_3-\overset{\displaystyle CH_3}{\underset{\displaystyle CH_3}{\overset{|}{\underset{|}{C}}}}-Br \quad \underset{}{\overset{slow}{\rightleftharpoons}} \quad CH_3-\overset{\displaystyle CH_3}{\underset{\displaystyle CH_3}{\overset{|}{\underset{|}{C^+}}}} \; + \; :Br^-$$

This reaction is possible because tertiary carbocations are relatively stable compared with secondary or primary ones. Even so, the reaction is slow.

Once the carbocation is formed, however, it would react immediately it came into contact with a nucleophile like Nu^-. The lone pair on the nucleophile is strongly attracted towards the positive carbon, and moves towards it to create a new bond.

$$CH_3-\overset{CH_3}{\underset{CH_3}{C^+}} \;\; :\bar{N}u \xrightarrow{\text{fast}} CH_3-\overset{CH_3}{\underset{CH_3}{C}}-Nu$$

How fast the reaction happens is going to be governed by how fast the halogenoalkane ionizes. Because this initial slow step only involves one species, the mechanism is described as S_N1 - substitution, nucleophilic, one species taking part in the initial slow step.

Why don't Primary Halogenoalkanes Use the S_N1 Mechanism?

If a primary halogenoalkane did use this mechanism, the first step would be, for example:

$$CH_3-CH_2-Br \overset{\text{slow}}{\rightleftharpoons} CH_3-\overset{+}{C}H_2 + :Br^-$$

A primary carbocation would be formed, and this is much more energetically unstable than the tertiary one formed from tertiary halogenoalkanes - and therefore much more difficult to produce.

This instability means that there will be a very high activation energy for the reaction involving a primary halogenoalkane. The activation energy is much less if it undergoes an S_N2 reaction - and so that's what it does instead.

Nucleophilic Substitution in Secondary Halogenoalkanes

$$CH_3-\underset{Br}{\underset{|}{C}H}-CH_3$$

a secondary halogenoalkane

There isn't anything new in this. Secondary halogenoalkanes will use both mechanisms - some molecules will react using the S_N2 mechanism and others the S_N1.

The S_N2 mechanism is possible because the back of the molecule isn't completely cluttered by alkyl groups and so the approaching nucleophile can still get at the $\delta+$ carbon atom.

The S_N1 mechanism is possible because the secondary carbocation formed in the slow step is more stable than a primary one. It isn't as stable as a tertiary one though, and so the S_N1 route isn't as effective as it is with tertiary halogenoalkanes.

Nucleophilic substitution reactions between halogenoalkanes and hydroxide ions

Reaction of Primary Halogenoalkanes with Hydroxide Ions

If a halogenoalkane is heated under reflux with a solution of sodium or potassium hydroxide, the halogen is replaced by -OH and an alcohol is produced. Heating under reflux means heating with a condenser placed vertically in the flask to prevent loss of volatile substances from the mixture.

The solvent is usually a 50/50 mixture of ethanol and water, because everything will dissolve in that. The halogenoalkane is insoluble in water. If you used water alone as the solvent, the halogenoalkane and the sodium hydroxide solution wouldn't mix and the reaction could only happen where the two layers met.

For example, using 1-bromopropane as a typical primary halogenoalkane:

$$CH_3CH_2CH_2Br + OH^- \longrightarrow CH_3CH_2CH_2OH + Br$$

You could write the full equation rather than the ionic one, but it slightly obscures what's going on:

$$CH_3CH_2CH_2Br + NaOH \longrightarrow CH_3CH_2CH_2OH + NaBr$$

The bromine (or other halogen) in the halogenoalkane is simply replaced by an -OH group - hence a substitution reaction. In this example, propan-1-ol is formed.

Mechanism

Here is the mechanism for the reaction involving bromoethane:

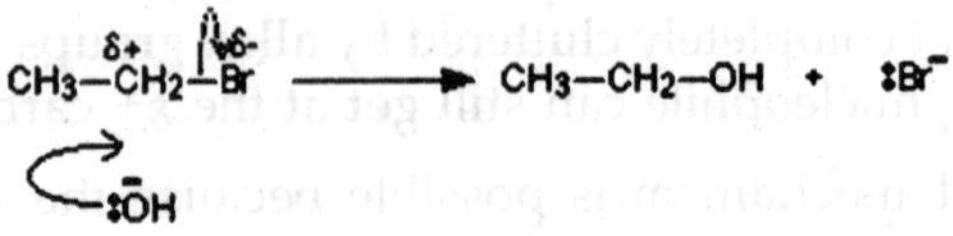

$$CH_3{-}CH_2{-}Br \longrightarrow CH_3{-}CH_2{-}OH + :Br^-$$

:OH⁻

This is an example of nucleophilic substitution

Because the mechanism involves collision between two species in the slow step (in this case, the only step) of the reaction, it is known as an S_N2 reaction

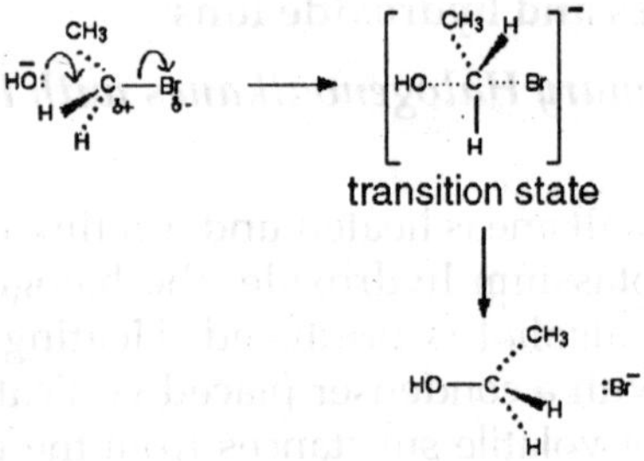

transition state

Reaction of Tertiary Halogenoalkanes with Hydroxide Ions

The facts of the reaction are exactly the same as with primary halogenoalkanes. If the halogenoalkane is heated under reflux with a solution of sodium or potassium hydroxide in a mixture of ethanol and water, the halogen is replaced by -OH, and an alcohol is produced.

For example:

$$(CH_3)_3CBr + OH^- \longrightarrow (CH_3)_3COH + Br^-$$

Or if you want the full equation rather than the ionic one:

$$(CH_3)_3CBr + NaOH \longrightarrow (CH_3)_3COH + NaBr$$

Mechanism

This mechanism involves an initial ionization of the halogenoalkane:

This mechanism involves an initial ionization of the halogenoalkane:

$$CH_3-\underset{CH_3}{\overset{CH_3}{C}}-Br \xrightleftharpoons{\text{slow}} CH_3-\underset{CH_3}{\overset{CH_3}{C^+}} + :Br^-$$

followed by a very rapid attack by the hydroxide ion on the carbocation (carbonium ion) formed:

$$CH_3-\underset{CH_3}{\overset{CH_3}{C^+}} \;\; :\bar{O}H \xrightarrow{\text{fast}} CH_3-\underset{CH_3}{\overset{CH_3}{C}}-OH$$

This reaction proceeds by nucleophilic substitution.

This time the slow step of the reaction only involves one species - the halogenoalkane. It is known as an S_N1 reaction.

Reaction of Secondary Halogenoalkanes with Hydroxide Ions

The facts of the reaction are exactly the same as with primary or tertiary halogenoalkanes. The halogenoalkane is heated under reflux with a solution of sodium or potassium hydroxide in a mixture of ethanol and water.

For example:

$$CH_3\underset{Br}{CH}CH_3 + OH^- \longrightarrow CH_3\underset{OH}{CH}CH_3 + Br^-$$

Mechanism

Secondary halogenoalkanes use *both* S_N2 and S_N1 mechanisms. For example, the S_N2 mechanism is:

$$\overset{:OH^-}{} \;\; CH_3-\underset{Br^{\delta-}}{\overset{\delta+}{CH}}-CH_3 \longrightarrow CH_3-\overset{OH}{CH}-CH_3 \quad :Br^-$$

The two stages of the S_N1 mechanism are:

$$CH_3-\underset{Br}{CH}-CH_3 \xrightleftharpoons{\text{slow}} CH_3-\overset{+}{C}H-CH_3 + :Br^-$$

$$CH_3-\overset{+}{C}H-CH_3 + {}^{-}\!:OH \xrightarrow{\text{fast}} CH_3-\underset{}{\overset{OH}{\overset{|}{C}H}}-CH_3$$

Nucleophilic Substitution Reactions between Halogenoalkanes and Water

Reaction of Primary Halogenoalkanes with Water

There is only a slow reaction between a primary halogenoalkane and water even if they are heated. The halogen atom is replaced by -OH.

For example, using 1-bromoethane as a typical primary halogenoalkane:

$$CH_3CH_2Br + H_2O \longrightarrow CH_3CH_2OH + HBr$$

An alcohol is produced together with hydrobromic acid. Be careful not to call this hydrogen bromide. Hydrogen bromide is a gas. When it is dissolved it in water (as it will be here), it's called hydrobromic acid.

Mechanism

The mechanism involves two steps. The first is a simple nucleophilic substitution reaction:

$$CH_3-\overset{\delta+}{C}H_2-\overset{\delta-}{Br} + H-\ddot{O}:-H \longrightarrow CH_3-CH_2-\overset{+}{O}H_2 + :Br^-$$

Because the mechanism involves collision between two species in this slow step of the reaction, it is known as an S_N2 reaction.

The nucleophilic substitution is very slow because water isn't a very good nucleophile. It lacks the full negative charge of, say, a hydroxide ion.

The second step of the reaction simply tidies up the product. A water molecule removes one of the hydrogens attached to the oxygen to give an alcohol and a hydroxonium ion (also known as a hydronium ion or an oxonium ion).

$CH_3-CH_2-\overset{+}{O}(H)-H + H_2\ddot{O} \longrightarrow CH_3-CH_2-\ddot{O}-H + H-\overset{+}{\ddot{O}}(H)-H$

The hydroxonium ion and the bromide ion (from the nucleophilic substitution stage of the reaction) make up the hydrobromic acid which is formed as well as the alcohol.

Reaction of Tertiary Halogenoalkanes with Water

If the halogenoalkane is heated under reflux with water, the halogen is replaced by -OH to give an alcohol. Heating under reflux means heating with a condenser placed vertically in the flask to prevent loss of volatile substances from the mixture. The reaction happens much faster than the corresponding one involving a primary halogenoalkane.

For example:

$$(CH_3)_3CBr + H_2O \longrightarrow (CH_3)_3COH + HBr$$

Mechanism

This mechanism involves an initial ionization of the halogenoalkane:

$$(CH_3)_3C-Br \underset{}{\overset{\text{slow}}{\rightleftharpoons}} (CH_3)_3C^+ + :Br^-$$

followed by a very rapid attack by the water on the carbocation (carbonium ion) formed:

$$(CH_3)_3C^+ + :\ddot{O}H_2 \xrightarrow{\text{fast}} (CH_3)_3C-\overset{+}{\ddot{O}}H_2$$

This proceeds by *nucleophilic substitution.*

This time the slow step of the reaction only involves one species - the halogenoalkane. It is known as an S_N1 reaction.

Now there is a final stage in which the product is tidied up. A water molecule removes one of the hydrogens attached to the oxygen to give an alcohol and a hydroxonium ion - exactly as happens with primary halogenoalkanes.

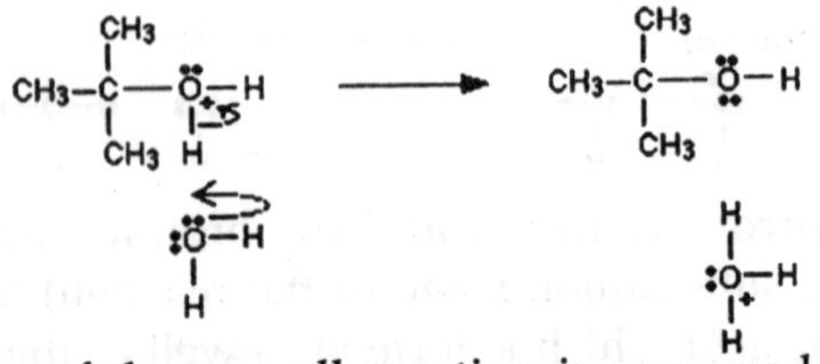

$$(CH_3)_2C(CH_3)-\overset{+}{O}H_2 + H_2O \longrightarrow (CH_3)_3C-OH + H_3O^+$$

The rate of the overall reaction is governed entirely by how fast the halogenoalkane ionizes. The fact that water isn't as good a nucleophile as, say, OH^- doesn't make any difference. The water isn't involved in the slow step of the reaction.

Nucleophilic substitution reactions between halogenoalkanes and cyanide ions

Reaction of Primary Halogenoalkanes with Cyanide Ions

If a halogenoalkane is heated under reflux with a solution of sodium or potassium cyanide in ethanol, the halogen is replaced by a -CN group and a nitrile is produced. Heating under reflux means heating with a condenser placed vertically in the flask to prevent loss of volatile substances from the mixture.The solvent is important. If water is present you tend to get substitution by -OH instead of -CN.

For example, using 1-bromopropane as a typical primary halogenoalkane:

$$CH_3CH_2CH_2Br + CN^- \longrightarrow CH_3CH_2CH_2CN + Br$$

You could write the full equation rather than the ionic one, but it slightly obscures what's going on:

$$CH_3CH_2CH_2Br + KCN \longrightarrow CH_3CH_2CH_2CN + KBr$$

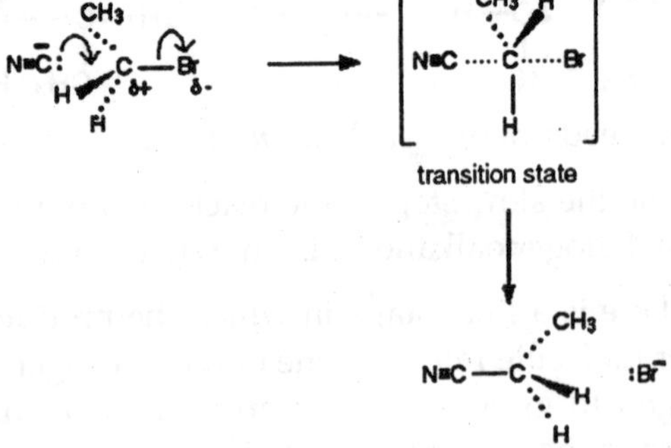

$$N{\equiv}C:^- + CH_3CH_2Br \longrightarrow [N{\equiv}C\cdots C(CH_3)(H)(H)\cdots Br]^- \longrightarrow N{\equiv}C-CH_2CH_3 + :Br^-$$

transition state

The bromine (or other halogen) in the halogenoalkane is simply replaced by a -CN group - hence a substitution reaction. In this example, butanenitrile is formed.

Reaction of Tertiary Halogenoalkanes with Cyanide Ions

The facts of the reaction are exactly the same as with primary halogenoalkanes. If the halogenoalkane is heated under reflux with a solution of sodium or potassium cyanide in ethanol, the halogen is replaced by -CN, and a nitrile is produced.

For example:

$$(CH_3)_3CBr + CN^- \longrightarrow (CH_3)_3CCN + Br$$

Or if you want the full equation rather than the ionic one:

$$(CH_3)_3CBr + KCN \longrightarrow (CH_3)_3CCN + KBr$$

Mechanism

This mechanism involves an initial ionization of the halogenoalkane:

$$CH_3-C(CH_3)_2-Br \underset{}{\overset{slow}{\rightleftharpoons}} CH_3-C^+(CH_3)_2 + :Br^-$$

followed by a very rapid attack by the cyanide ion on the carbocation (carbonium ion) formed:

$$CH_3-C^+(CH_3)_2 \;\; :\bar{C}{\equiv}N \xrightarrow{fast} CH_3-C(CH_3)_2-C{\equiv}N$$

This time the slow step of the reaction only involves one species - the halogenoalkane. It is known as an S_N1 reaction.

Reaction of Secondary Halogenoalkanes with Cyanide Ions

The facts of the reaction are exactly the same as with primary or tertiary halogenoalkanes. The halogenoalkane is heated under reflux with a solution of sodium or potassium cyanide in ethanol.

For example:

$$\underset{\displaystyle |\atop Br}{CH_3CHCH_3} + CN^- \longrightarrow \underset{\displaystyle |\atop CN}{CH_3CHCH_3} + Br^-$$

Nucleophilic Addition / Elimination in the Reactions of Acyl Chlorides

Acyl chlorides (also known as acid chlorides) are one of a number of types of compounds known as "acid derivatives". This is ethanoic acid:

$$CH_3-C(=O)-OH$$

If you remove the -OH group and replace it by a -Cl, you have produced an acyl chloride.

$$CH_3-C(=O)-Cl$$

This molecule is known as ethanoyl chloride and for the rest of this topic will be taken as typical of acyl chlorides in general.

Acyl chlorides are extremely reactive. They are open to attack by nucleophiles - with the overall result being a replacement of the chlorine by something else.

Why are Acyl chlorides Attacked by Nucleophiles?

The carbon atom in the -COCl group has both an oxygen atom and a chlorine atom attached to it. Both of these are very electronegative. They both pull electrons towards themselves, leaving the carbon atom quite positively charged.

$$CH_3-\overset{\delta+}{C}(=O^{\delta-})-Cl^{\delta-}$$

easy attack by a nucleophile

The Overall Reaction

We are going to generalize this for the moment by writing the reacting molecule as "Nu-H". Nu is the bit of the molecule which contains the nucleophilic oxygen or nitrogen atom. The attached hydrogen turns out to be essential to the reaction.

The general equation for the reaction is:

$$CH_3{-}C({=}O){-}Cl + Nu{-}H \longrightarrow CH_3{-}C({=}O){-}Nu + HCl$$

In each case, the net effect is that you replace the -Cl by -Nu, and hydrogen chloride is formed as well.

Since the initial attack is by a nucleophile, and the overall result is substitution, it would seem reasonable to describe the reaction as nucleophilic substitution. However, the reaction happens in two distinct stages. The first involves an addition reaction, which is followed by an elimination reaction where HCl is produced. So the mechanism is also known as nucleophilic addition / elimination.

General mechanism

The Addition Stage of the Mechanism

As the lone pair on the nucleophile approaches the fairly positive carbon in the acyl chloride, it moves to form a bond with it. In the process, the two electrons in one of the carbon-oxygen bonds are repelled entirely onto the oxygen, leaving that oxygen negatively charged.

$$CH_3{-}C^{\delta+}({=}O^{\delta-}){-}Cl^{\delta-} + \ddot{N}u{-}H \longrightarrow CH_3{-}C(O^-)(Cl){-}Nu^+{-}H$$

Notice the positive charge that forms on the nucleophile. Just accept this for the moment. It's much easier to explain why that charge must be there if you have a real example in front of you.

The Elimination Stage of the Mechanism

This happens in two steps. In the first step, the carbon-oxygen double bond reforms. To make room for it, the electrons in the carbon-chlorine bond are repelled until they are entirely on the chlorine atom - forming a chloride ion.

$$CH_3-C(O^-)(Cl)(Nu^+H) \longrightarrow CH_3-C(=O)-Nu^+H \quad :Cl^-$$

Finally, the chloride ion plucks the hydrogen off the original nucleophile. It removes it as a hydrogen ion, leaving the pair of electrons behind on the oxygen or nitrogen atom in that nucleophile. That cancels the positive charge.

$$CH_3-C(=O)-Nu^+-H \quad :Cl^- \longrightarrow CH_3-C(=O)-Nu \quad HCl$$

Elimination Reactions Producing Alkenes from Simple Halogenoalkanes

The Elimination Reaction Involving 2-bromopropane and Hydroxide ions

2-bromopropane is heated under reflux with a concentrated solution of sodium or potassium hydroxide in ethanol. Heating under reflux involves heating with a condenser placed vertically in the flask to avoid loss of volatile liquids. Propene is formed and, because this is a gas, it passes through the condenser and can be collected.

$$CH_3CH(Br)CH_3 + NaOH \longrightarrow CH_2{=}CHCH_3 + NaBr + H_2O$$

Everything else present (including anything formed in the alternative substitution reaction) will be trapped in the flask.

Mechanism

In elimination reactions, the hydroxide ion acts as a base - removing a hydrogen as a hydrogen ion from the carbon atom next door to the one holding the bromine.

The resulting re-arrangement of the electrons expels the bromine as a bromide ion and produces propene.

$:\bar{O}H$ H_2O

$H-\underset{H}{\overset{H}{C}}-\underset{Br}{CH}-CH_3 \longrightarrow H-\underset{H}{C}=CHCH_3$

$:Br^-$

Elimination from Unsymmetric Halogenoalkanes

$CH_3-\underset{Br}{CH}-CH_2-CH_3$

2-bromobutane is an unsymmetric halogenoalkane in the sense that it has a CH_3 group one side of the C-Br bond and a CH_2CH_3 group the other.

You have to be careful with compounds like this because of the possibility of more than one elimination product depending on where the hydrogen is removed from.

The basic facts and mechanisms for these reactions are exactly the same as with simple halogenoalkanes like 2-bromopropane.

You will remember that elimination happens when a hydroxide ion (from, for example, sodium hydroxide) acts as a base and removes a hydrogen as a hydrogen ion from the halogenoalkane.

For example, in the simple case of elimination from 2-bromopropane:

$:\bar{O}H$ H_2O

$H-\underset{H}{\overset{H}{C}}-\underset{Br}{CH}-CH_3 \longrightarrow H-\underset{H}{C}=CHCH_3$

$:Br^-$

The hydroxide ion removes a hydrogen from one of the carbon atoms next door to the carbon-bromine bond, and the various electron shifts then lead to the formation of the alkene - in this case, propene.

With an unsymmetric halogenoalkane like 2-bromobutane, there are several hydrogens which might possibly get removed. You need to think about each of these possibilities.

Where does the Hydrogen get Removed from?

The hydrogen has to be removed from a carbon atom adjacent to the carbon-bromine bond. If an OH^- ion hit one of the hydrogens on the right-hand CH_3 group in the 2-bromobutane (as we've drawn it), there's nowhere for the reaction to go.

$$CH_3-CH(Br)-CH_2-CH_3 + {:}\bar{O}H$$

This electron pair movement can't happen...

...because this movement would cost too much energy.

To make room for the electron pair to form a double bond between the carbons, you would have to expel a hydrogen from the CH_2 group as a hydride ion, H^-. That is energetically much too difficult, and so this reaction doesn't happen.

That still leaves the possibility of removing a hydrogen either from the left-hand CH_3 or from the CH_2 group.

If it was Removed from the CH_3 Group

$$H-CH_2-CH(Br)-CH_2-CH_3 + {:}\bar{O}H \longrightarrow H-CH=CH-CH_2-CH_3 + H_2O + {:}Br^-$$

The product is but-1-ene, $CH_2{=}CHCH_2CH_3$.

If it was Removed from the CH_2 Group:

$$CH_3-CH(Br)-CH_2-CH_3 + {:}\bar{O}H \longrightarrow CH_3-CH=CH-CH_3 + H_2O + {:}Br^-$$

This time the product is but-2-ene, $CH_3CH{=}CHCH_3$.

In fact the situation is even more complicated than it looks, because but-2-ene exhibits geometric isomerism. You get a mixture of two isomers formed - *cis*-but-2-ene and *trans*-but-2-ene.

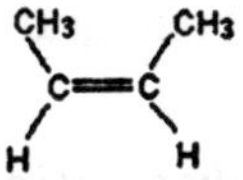

$$\underset{cis\text{-but-2-ene}}{CH_3(H)C{=}C(H)CH_3} \qquad \underset{trans\text{-but-2-ene}}{CH_3(H)C{=}C(CH_3)H}$$

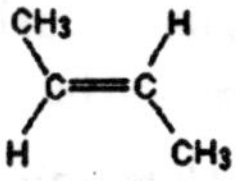

Which isomer gets formed is just a matter of chance.

The Overall Result

Elimination from 2-bromobutane leads to a mixture containing:

- But-1-ene
- Cis-but-2-ene
- Trans-but-2-ene

Elimination Versus Substitution in Halogenoalkanes

Both reactions involve heating the halogenoalkane under reflux with sodium or potassium hydroxide solution.

Nucleophilic Substitution

The hydroxide ions present are good nucleophiles, and one possibility is a replacement of the halogen atom by an -OH group to give an alcohol via a nucleophilic substitution reaction.

$$CH_3CH(Br)CH_3 + NaOH \longrightarrow CH_3CH(OH)CH_3 + NaBr$$

In the example, 2-bromopropane is converted into propan-2-ol.

Elimination

Halogenoalkanes also undergo elimination reactions in the presence of sodium or potassium hydroxide.

$$CH_3CH(Br)CH_3 + NaOH \longrightarrow CH_2{=}CHCH_3 + NaBr + H_2O$$

The 2-bromopropane has reacted to give an alkene - propene.

Notice that a hydrogen atom has been removed from one of the end carbon atoms together with the bromine from the centre one. In all simple elimination reactions the things being removed are on adjacent carbon atoms, and a double bond is set up between those carbons.

What Decides whether you get Substitution or Elimination?

The reagents you are using are the same for both substitution or elimination - the halogenoalkane and either sodium or potassium hydroxide solution. In all cases, you will get a mixture of both reactions happening - some substitution and some elimination. What you get *most* of depends on a number of factors.

The Type of Halogenoalkane

This is the most important factor.

Type of Halogenoalkane	*Substitution or Elimination?*
Primary	Mainly substitution
Secondary	Both substitution and elimination
Tertiary	Mainly elimination

For example, whatever you do with tertiary halogenoalkanes, you will tend to get mainly the elimination reaction, whereas with primary ones you will tend to get mainly substitution. However, you can influence things to some extent by changing the conditions.

The Solvent

The proportion of water to ethanol in the solvent matters.

- Water encourages substitution.
- Ethanol encourages elimination.

The Temperature

Higher temperatures encourage elimination.

Concentration of the Sodium or Potassium Hydroxide Solution

Higher concentrations favour elimination.

In Summary

For a given halogenoalkane, to favour elimination rather than substitution, use:

- Heat
- A concentrated solution of sodium or potassium hydroxide
- Pure ethanol as the solvent

Role of the Hydroxide Ion in a Substitution Reaction

In the substitution reaction between a halogenoalkane and OH^- ions, the hydroxide ions are acting as nucleophiles. For example, one of the lone pairs on the oxygen can attack the slightly positive carbon. This leads on to the loss of the bromine as a bromide ion, and the -OH group becoming attached in its place.

$$\bar{:OH} \rightarrow CH_3-\overset{\delta+}{C}H(Br^{\delta-})-CH_3$$

Role of the Hydroxide Ion in an Elimination Reaction

Hydroxide ions have a very strong tendency to combine with hydrogen ions to make water - in other words, the OH^- ion is a very strong base. In an elimination reaction, the hydroxide ion hits one of the hydrogen atoms in the CH_3 group and pulls it off. This leads to a cascade of electron pair movements resulting in the formation of a carbon-carbon double bond, and the loss of the bromine as Br

$$\bar{:OH} \rightarrow H-CH_2-CH(Br)-CH_3$$

Chapter 5

Alkanes

INTRODUCTION

Alkanes are the simplest family of hydrocarbons - compounds containing carbon and hydrogen only. They only contain carbon-hydrogen bonds and carbon-carbon single bonds. The general molecular formula for the members of the alkane family is:

C_nH_{2n+2}

where n = number of carbon atoms in the alkane molecule

The first ten normal members of the alkane family have n values of 1-10. A normal hydrocarbon alkane is one where all the carbon atoms in the molecule are in a "continuous" chain. A continuous chain does not necessarily mean that the carbons are in a straight line.

Normal Alkanes

Molecular Formula	*Structural Formula*	*IUPAC Name*
CH_4	CH_4	Methane
C_2H_6	CH_3-CH_3	Ethane
C_3H_8	CH_3-CH_2-CH_3	Propane
C_4H_{10}	CH_3-$(CH_2)_2$-CH_3	Butane
C_5H_{12}	CH_3-$(CH_2)_3$-CH_3	Pentane
C_6H_{14}	CH_3-$(CH_2)_4$-CH_3	Hexane
C_7H_{16}	CH_3-$(CH_2)_5$-CH_3	Heptane

C_8H_{18}	$CH_3\text{-}(CH_2)_6\text{-}CH_3$	Octane
C_9H_{20}	$CH_3\text{-}(CH_2)_7\text{-}CH_3$	Nonane
$C_{10}H_{22}$	$CH_3\text{-}(CH_2)_8\text{-}CH_3$	Decane

Notice that all the alkanes above end in "ane". This is considered the official IUPAC ending for alkanes.

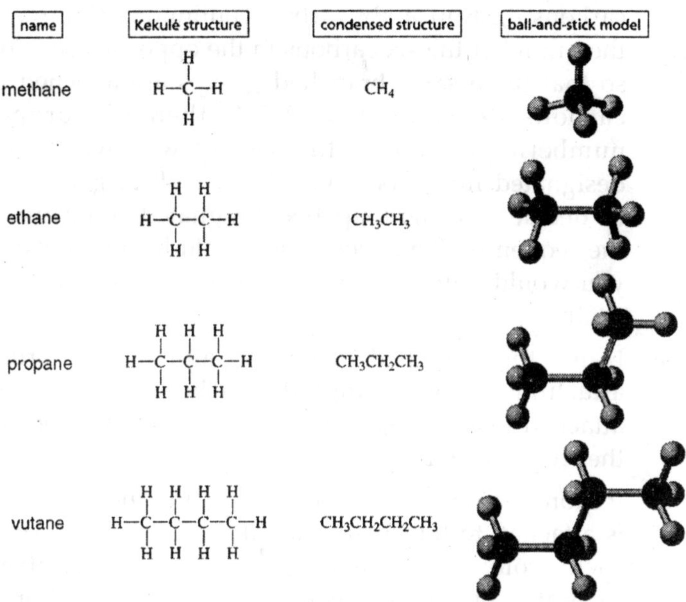

Fig. Few of the normal alkanes

IUPAC Rules For Naming Alkanes

1. Determine the longest continuous chain of carbons that have the functional group attached to one of the carbons. By "longest continuous chain" is meant to be able to trace through the carbons without raising the tracer (or finger) off the surface. The chain does not necessarily have to be straight.
2. Identify the various branching groups attached to this continuous chain of carbons by name
3. Number the carbons in the chain so that the groups would be attached to the carbons with the lowest designated number. This means that you have to

decide whether to number beginning on the right end or left end of the chain. The answer to that question will depend upon what the sequence numbers will be that establish the location of the branched groups. For example, if you number one way so that three branched groups are attached on number designated carbons 3,5,6 in a six carbon continuous chain and then number the six carbons in the opposite direction so that these same branched groups are attached to carbons designated as 1,2,4, then the proper numbering would be the second way where the designated numbers would be 1, 2, 4. A good way to decide is to add up the designated numbers in the sequence. The lowest sum would be the direction you would number of the carbons in the continuous chain.

4. Name the branched groups in alphabetical order attaching (hyphenating) the carbon number it is attached to along the continuous chain of carbons to the front of the branch name.

5. If more than one of the same kind of branched group is attached to the chain, identify the number carbon each group is attached to as a series of numbers separated by commas between each number then a hyphen and finally use a greek prefix attached to the branch name. For example, if there were three methyl groups (CH_3) attached to the 2,3,and 5 carbons along the chain, then that part of the name would be

   2,3,5-trimethyl.

   We use the prefix "tri" because there are three methyl groups

The greek prefixes used are as follows:

- Di for 2
- Tri for three
- Tetra for four
- Penta for five

- Hexa for six
- Hepta for seven
- Octa for eight

Identify the carbon number where the functional group is located or attached

End the name by attaching the normal alkane name corresponding to the number of carbon atoms in the normal alkane.

Isomerism

All the alkanes with 4 or more carbon atoms in them show structural isomerism. This means that there are two or more different structural formulae that you can draw for each molecular formula.

For example, C_4H_{10} could be either of these two different molecules:

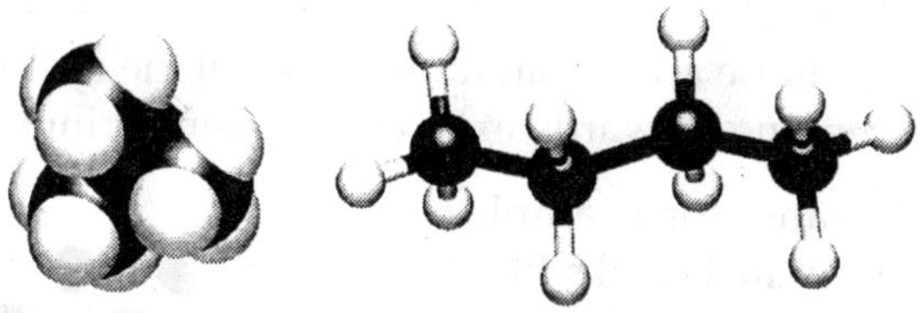

Fig. Butane **Fig.** 2-methylpropane

These are called respectively *butane* and *2-methylpropane.*

Preparations of Alkanes

Alkanes can be prepared by catalytic hydrogenation of alkenes or alkynes. For example:

1. $CH_2{=}CH_2 + H_2 \xrightarrow{Pt} CH_3{-}CH_3$
2. Alkanes can be prepared by a coupling reaction known as the Wurtz Reaction:

   $2CH_3{-}Br + 2Na^{\sim} \rightarrow CH_3{-}CH_3 + 2NaBr$
3. Alkanes can be prepared by forming an organo-metallic reagent known as a Grignard from an alkyl halide and then hydrolyzing it (react it with water):

4. $CH_3\text{-}Br + Mg \xrightarrow{\text{anhyd ether}} CH_3^- {}^+MgBr$

   $CH_3^- {}^+MgBr + H_2O \rightarrow CH_4 + MgBrOH$

5. Thermal Decarboxylation of a Carboxylate Salt

   $CH_3\text{-}CH_2\text{-}COOH + NaOH^- \rightarrow CH_3\text{-}CH_3 + CO_2$

Cycloalkanes

Cycloalkanes again only contain carbon-hydrogen bonds and carbon-carbon single bonds, but this time the carbon atoms are joined up in a ring. The smallest cycloalkane is cyclopropane.

Cycloalkanes

If you count the carbons and hydrogens, you will see that they no longer fit the general formula C_nH_{2n+2}. By joining the carbon atoms in a ring, you have had to lose two hydrogen atoms.

Thus the general formula for a cycloalkane is C_nH_{2n}.

Not all of these are all flat molecules. All the cycloalkanes from cyclopentane upwards exist as "puckered rings".

Cyclohexane, for example, has a ring structure which looks like this:

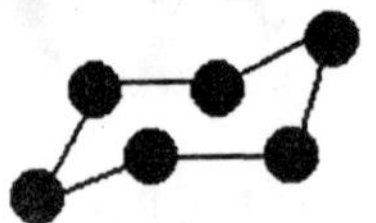

This is known as the "chair" form of cyclohexane - from its shape which vaguely resembles a chair.

Bonding in Methane and Ethane Methane, CH_4

The Simple View of the Bonding in Methane

You will be familiar with drawing methane using dots and crosses diagrams, but it is worth looking at its structure a bit more closely.

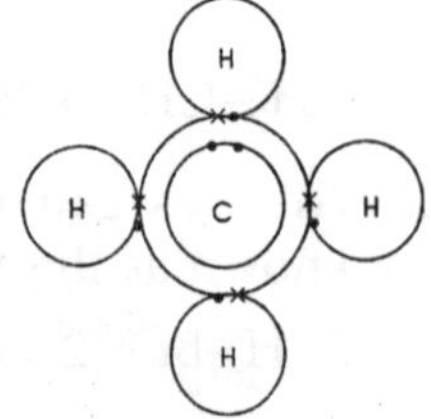

There is a serious mis-match between this structure and the modern electronic structure of carbon, $1s^22s^22p_x^{\ 1}2p_y^{\ 1}$. The modern structure shows that there are only 2 unpaired electrons for hydrogens to share with, instead of the 4 which the simple view requires.

You can see this more readily using the electrons-in-boxes notation. Only the 2-level electrons are shown. The $1s^2$ electrons are too deep inside the atom to be involved in bonding. The only electrons directly available for sharing are the 2p electrons. Why then isn't methane CH_2?

Promotion of an Electron

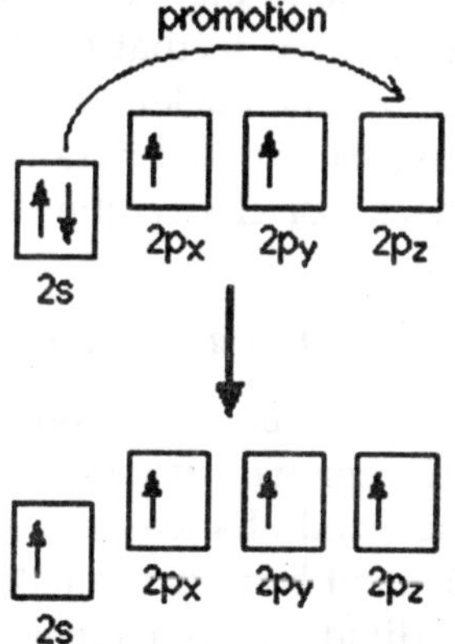

When bonds are formed, energy is released and the system becomes more stable. If carbon forms 4 bonds rather than 2, twice as much energy is released and so the resulting molecule becomes even more stable.

There is only a small energy gap between the 2s and 2p orbitals, and so it pays the carbon to provide a small amount of energy to promote an electron from the 2s to the empty 2p to give 4 unpaired electrons. The extra energy released when the bonds form more than compensates for the initial input.

Now that we've got 4 unpaired electrons ready for bonding, another problem arises. In methane all the carbon-hydrogen bonds are identical, but our electrons are in two different kinds of orbitals. You aren't going to get four identical bonds unless you start from four identical orbitals.

Hybridisation

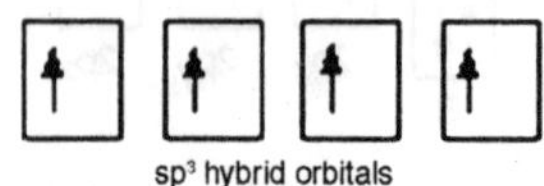

The electrons rearrange themselves again in a process called hybridisation. This reorganises the electrons into four identical hybrid orbitals called sp^3 hybrids (because they are made from one s orbital and three p orbitals). You should read "sp^3" as "s p three" - not as "s p cubed".

sp^3 hybrid orbitals look a bit like half a p orbital, and they arrange themselves in space so that they are as far apart as possible. You can picture the nucleus as being at the centre of a tetrahedron (a triangularly based pyramid) with the orbitals pointing to the corners. For clarity, the nucleus is drawn far larger than it really is.

What Happens when the Bonds are Formed?

Remember that hydrogen's electron is in a 1s orbital - a spherically symmetric region of space surrounding the nucleus where there is some fixed chance (say 95%) of finding the electron. When a covalent bond is formed, the atomic orbitals (the orbitals in the individual atoms) merge to produce a new molecular orbital which contains the electron pair which creates the bond.

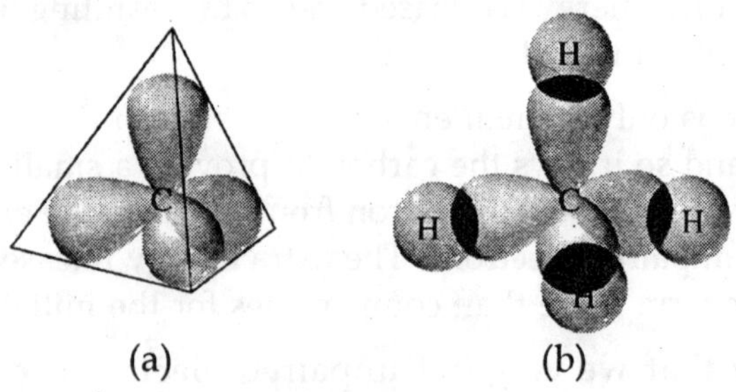

Fig. Bonding in Methane

Four molecular orbitals are formed, looking rather like the original sp^3 hybrids, but with a hydrogen nucleus embedded in each lobe. Each orbital holds the 2 electrons that we've previously drawn as a dot and a cross.

The principles involved - promotion of electrons if necessary, then hybridisation, followed by the formation of molecular orbitals - can be applied to any covalently-bound molecule.

Ethane, C_2H_6

The Formation of Molecular Orbitals in Ethane

Ethane isn't particularly important in its own right, but is included because it is a simple example of how a carbon-carbon single bond is formed.

Each carbon atom in the ethane promotes an electron and then forms sp^3 hybrids exactly as we've described in methane. So just before bonding, the atoms look like this:

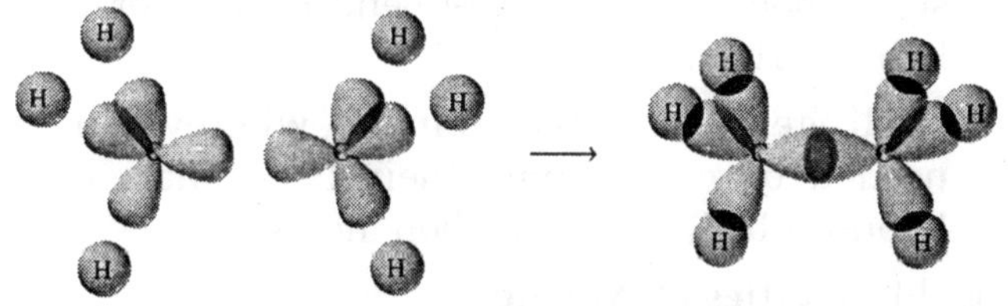

Fig. Bond formation in ethane

The hydrogens bond with the two carbons to produce molecular orbitals just as they did with methane. The two carbon atoms bond by merging their remaining sp^3 hybrid orbitals end-to-end to make a new molecular orbital. The bond formed by this end-to-end overlap is called a *sigma bond.* The bonds between the carbons and hydrogens are also sigma bonds.

In any sigma bond, the most likely place to find the pair of electrons is on a line between the two nuclei.

Free Rotation about the Carbon-Carbon Single Bond

The two ends of this molecule can spin quite freely about the sigma bond so that there are, in a sense, an infinite number of possibilities for the shape of an ethane molecule. Some possible shapes are:

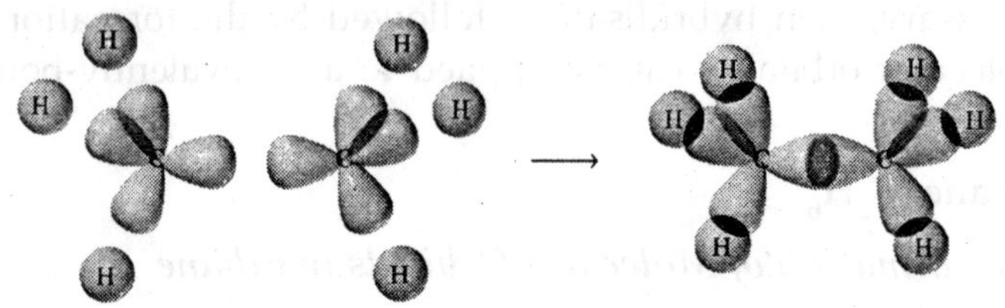

Fig. Rotation about C-C bond

Other Alkanes

All other alkanes will be bonded in the same way:

- The carbon atoms will each promote an electron and then hybridise to give sp^3 hybrid orbitals.
- The carbon atoms will join to each other by forming sigma bonds by the end-to-end overlap of their sp^3 hybrid orbitals.
- Hydrogen atoms will join on wherever they are needed by overlapping their $1s^1$ orbitals with sp^3 hybrid orbitals on the carbon atoms.

Physical Properties of Alkanes

Boiling Points

The boiling points of organic compounds are affected by molecular size and the degree of intermolecular force that exists between the molecules. Generally speaking, the boiling points of alkanes are the lowest of any Organic Family if you compare roughly the same size molecules. This is because the intermolecular forces holding the alkane molecules together are very weak London Dispersion forces. These are the only intermolecular forces that would be present. As a result, the temperature required to separate the molecules into the vapor state is relatively low.

Branching tends to lower the boiling point since the branching decreases the surface area of the molecule where London Dispersion forces are most pronounced. Notice that the first four alkanes are gases at room temperature. Solids don't start to appear until about $C_{17}H_{36}$.

You can't be more precise than that because each isomer has a different melting and boiling point. By the time you get 17 carbons into an alkane, there are unbelievable numbers of isomers!

Cycloalkanes have boiling points which are about 10 - 20 K higher than the corresponding straight chain alkane.

Explanations

There isn't much electronegativity difference between carbon and hydrogen, so there is hardly any bond polarity. The molecules themselves also have very little polarity. A totally symmetrical molecule like methane is completely non-polar. This means that the only attractions between one molecule and its neighbours will be Van der Waals dispersion forces. These will be very small for a molecule like methane, but will increase as the molecules get bigger. That's why the boiling points of the alkanes increase with molecular size. Where you have isomers, the more branched the chain, the lower the boiling point tends to be. Van der Waals dispersion forces are smaller for shorter molecules, and only operate over very short distances between one molecule and its neighbours. It is more difficult for short fat molecules (with lots of branching) to lie as close together as long thin ones.

For example, the boiling points of the three isomers of C_5H_{12} are:

Alkane	*Boiling point (K)*
Pentane	309.2
2-methylbutane	301.0
2,2-dimethylpropane	282.6

The slightly higher boiling points for the cycloalkanes are presumably because the molecules can get closer together because the ring structure makes them tidier and less complicated.

Solubility

Alkanes are insoluble in water because the polar water molecules are not attracted to the non-polar alkane molecules. However alkanes are soluble in non-polar solvents such as Carbon Tetrachloride,CCl_4. Generally speaking like disolves in like and unlike do not disolve. So non-polar substances will be insoluble in polar solvents such as water, but they will be soluble in non-polar solvents (like dissolves in like).

The liquid alkanes are good solvents for many other covalent compounds.

Explanations

Solubility in water

When a molecular substance dissolves in water, you have to·

- Break the intermolecular forces within the substance. In the case of the alkanes, these are Van der Waals dispersion forces.
- Break the intermolecular forces in the water so that the substance can fit between the water molecules. In water the main intermolecular attractions are hydrogen bonds.

Breaking either of these attractions costs energy, although the amount of energy to break the Van der Waals dispersion forces in something like methane is pretty negligible. That isn't true of the hydrogen bonds in water, though.As something of a simplification, a substance will dissolve if there is enough energy released when new bonds are made between the substance and the water to make up for what is used in breaking the original attractions.The only new attractions between the alkane and water molecules are Van der Waals. These don't release anything like enough energy to compensate for what you need to break the hydrogen bonds in water. The alkane doesn't dissolve.

Solubility in Organic Solvents

In most organic solvents, the main forces of attraction between the solvent molecules are Van der Waals - either dispersion forces or dipole-dipole attractions.

That means that when an alkane dissolves in an organic solvent, you are breaking Van der Waals forces and replacing them by new Van der Waals forces. The two processes more or less cancel each other out energetically - so there isn't any barrier to solubility.

Reactivity of Alkanes and Cycloalkanes

Alkanes

Alkanes contain strong carbon-carbon single bonds and strong carbon-hydrogen bonds. The carbon-hydrogen bonds are only very slightly polar and so there aren't any bits of the molecules which carry any significant amount of positive or negative charge which other things might be attracted to.

The net effect is that alkanes have a fairly restricted set of reactions.

You can

- Burn them - destroying the whole molecule;
- React them with some of the halogens, breaking carbon-hydrogen bonds;
- Crack them, breaking carbon-carbon bonds.

Cycloalkanes

Cycloalkanes are very similar to the alkanes in reactivity, except for the very small ones - especially cyclopropane. Cyclopropane is much more reactive than you would expect.

The reason has to do with the bond angles in the ring. Normally, when carbon forms four single bonds, the bond angles are about 109.5°. In cyclopropane, they are 60°.

With the electron pairs this close together, there is a lot of repulsion between the bonding pairs joining the carbon atoms. That makes the bonds easier to break.

The effect of this is explored on the page about reactions of these compounds with halogens.

Reactions of Alkanes and Cycloalkanes

Combustion of Alkanes and Cycloalkanes

Complete Combustion

Complete combustion (given sufficient oxygen) of any hydrocarbon produces carbon dioxide and water.

For example, with propane (C_3H_8), we get:

$$C_3H_8 + 5O_2 \longrightarrow 3CO_2 + 4H_2)$$

With butane (C_4H_{10}), the equation is:

$$C_4H_{10} + 6\tfrac{1}{2}O_2 \longrightarrow 4CO_2 + 5H_2O$$

The hydrocarbons become harder to ignite as the molecules get bigger. This is because the bigger molecules don't vaporize so easily - the reaction is much better if the oxygen and the hydrocarbon are well mixed as gases. If the liquid isn't very volatile, only those molecules on the surface can react with the oxygen.

Bigger molecules have greater Van der Waals attractions which makes it more difficult for them to break away from their neighbours and turn to a gas.

Provided the combustion is complete, all the hydrocarbons will burn with a blue flame. However, combustion tends to be less complete as the number of carbon atoms in the molecules rises. That means that the bigger the hydrocarbon, the more likely you are to get a yellow, smoky flame.

Incomplete Combustion

Incomplete combustion (where there isn't enough oxygen present) can lead to the formation of carbon or carbon monoxide.

The presence of glowing carbon particles in a flame turns it yellow, and black carbon is often visible in the smoke.

Carbon monoxide is produced as a colourless poisonous gas.

Cracking

Cracking is the name given to breaking up large hydrocarbon molecules into smaller and more useful bits. This is achieved by using high pressures and temperatures without a catalyst, or lower temperatures and pressures in the presence of a catalyst.

The source of the large hydrocarbon molecules is often the naphtha fraction or the gas oil fraction from the fractional distillation of crude oil (petroleum). These fractions are obtained from the distillation process as liquids, but are re-vaporized before cracking.

There isn't any single unique reaction happening in the cracker. The hydrocarbon molecules are broken up in a fairly random way to produce mixtures of smaller hydrocarbons, some of which have carbon-carbon double bonds. One possible reaction involving the hydrocarbon $C_{15}H_{32}$ might be:

$$C_{15}H_{32} \longrightarrow \underset{\text{ethene}}{2C_2H_4} + \underset{\text{propene}}{C_3H_6} + \underset{\text{octane}}{C_8H_{18}}$$

The ethene and propene are important materials for making plastics or producing other organic chemicals. The octane is one of the molecules found in petrol (gasoline).

Catalytic Cracking

Modern cracking uses zeolites as the catalyst. These are complex aluminosilicates, and are large lattices of aluminium, silicon and oxygen atoms carrying a negative charge.

They are, of course, associated with positive ions such as sodium ions. You may have come across a zeolite if you know about ion exchange resins used in water softeners.

The alkane is brought into contact with the catalyst at a temperature of about 500°C and moderately low pressures.

The zeolites used in catalytic cracking are chosen to give high percentages of hydrocarbons with between 5 and 10 carbon atoms - particularly useful for petrol (gasoline). It also produces high proportions of branched alkanes and aromatic hydrocarbons like benzene.

The zeolite catalyst has sites which can remove a hydrogen from an alkane together with the two electrons which bound it to the carbon. That leaves the carbon atom with a positive charge. Ions like this are called carbonium ions (or carbocations). Reorganization of these leads to the various products of the reaction.

Halogenation

Alkanes

The Reaction between Alkanes and Fluorine

This reaction is explosive even in the cold and dark, and you tend to get carbon and hydrogen fluoride produced. It is of no particular interest. For example:

$$CH_4 + 2F_2 \longrightarrow C + 4HF$$

The Reaction between Alkanes and Iodine

Iodine doesn't react with the alkanes - at least, under normal lab conditions.

The Reactions between Alkanes and Chlorine or Bromine

There is no reaction in the dark.

In the presence of a flame, the reactions are rather like the fluorine one - producing a mixture of carbon and the hydrogen halide. The violence of the reaction drops considerably as you go from fluorine to chlorine to bromine.

The interesting reactions happen in the presence of ultraviolet light (sunlight will do). These are *photochemical reactions,* and happen at room temperature.

We'll look at the reactions with chlorine. The reactions with bromine are similar, but rather slower.

Examples:

Methane and Chlorine: Substitution reactions happen in which hydrogen atoms in the methane are replaced one at a time by chlorine atoms. You end up with a mixture of chloromethane, dichloromethane, trichloromethane and tetrachloromethane.

$$CH_4 + Cl_2 \longrightarrow CH_3Cl + HCl$$

chloromethane

$$CH_3Cl + Cl_2 \longrightarrow CH_2Cl_2 + HCl$$

cdichloromethane

$$CH_2Cl_2 + Cl_2 \longrightarrow CHCl_3 + HCl$$

trichloromethane

$$CHCl_3 + Cl_2 \longrightarrow CCl_4 + HCl$$

tetrachloromethane

The original mixture of a colourless and a green gas would produce steamy fumes of hydrogen chloride and a mist of organic liquids. All of the organic products are liquid at room temperature with the exception of the chloromethane which is a gas.

If you were using bromine, you could either mix methane with bromine vapour, or bubble the methane through liquid bromine - in either case, exposed to UV light. The original mixture of gases would, of course, be red-brown rather than green.

You shouldn't choose to use these reactions as a means of preparing these organic compounds in the lab because the mixture of products would be too tedious to separate.

Larger Alkanes and Chlorine: You would again get a mixture of substitution products, but it is worth just looking briefly at what happens if only one of the hydrogen atoms gets substituted (monosubstitution).

For example, with propane, you could get one of two isomers:

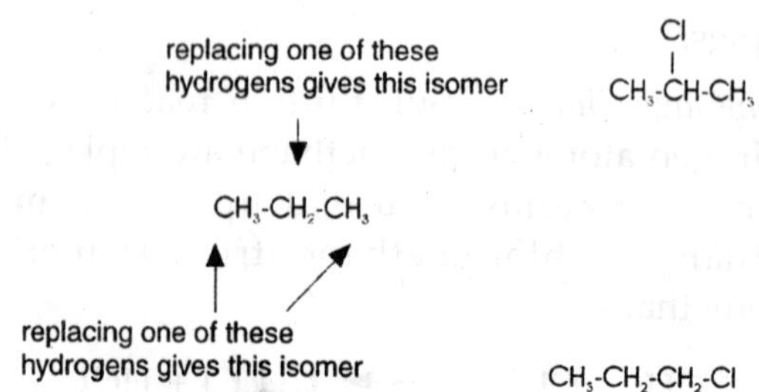

If chance was the only factor, you would expect to get 3 times as much of the isomer with the chlorine on the end. There are 6 hydrogens that could get replaced on the end carbon atoms compared with only 2 in the middle.

In fact, you get about the same amount of each of the two isomers.

If you use bromine instead of chlorine, the great majority of the product is where the bromine is attached to the centre carbon atom.

Cycloalkanes

The reactions of the cycloalkanes are generally just the same as the alkanes, with the exception of the very small ones - particularly cyclopropane.

The Extra Reactivity of Cyclopropane

In the presence of UV light, cyclopropane will undergo substitution reactions with chlorine or bromine just like a non-cyclic alkane. However, it also has the ability to react in the dark.

In the absence of UV light, cyclopropane can undergo addition reactions in which the ring is broken. For example, with bromine, cyclopropane gives 1,3-dibromopropane.

CH_2

H_2C — CH_2 $+ Br_2 \longrightarrow Br\text{-}CH_2\text{-}CH_2\text{-}CH_2\text{-}Br$

This can still happen in the presence of light - but you will get substitution reactions as well.

The ring is broken because cyclopropane suffers badly from ring strain. The bond angles in the ring are 60° rather than the normal value of about 109.5° when the carbon makes four single bonds.

The overlap between the atomic orbitals in forming the carbon-carbon bonds is less good than it is normally, and there is considerable repulsion between the bonding pairs. The system becomes more stable if the ring is broken.

Chapter 6

Alkene

Introduction

Alkenes are a family of hydrocarbons (compounds containing carbon and hydrogen only) containing a carbon-carbon double bond. The general formula for alkenes with only one double bond in the molecule is C_nH_{2n} where n is the number of carbon atoms in the molecule.

The first two are:

Ethene C_2H_4

Propene C_3H_6

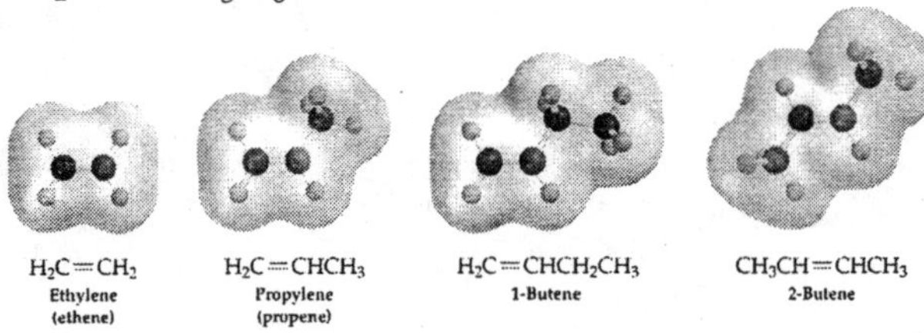

Fig. Some common alkenes

IUPAC Rules For Alkenes

1. Determine the longest continuous chain of carbons that have the double bond between two of its carbons. By "longest continuous chain" is meant to be able to trace through the carbons without raising the tracer (or finger) off the surface. The chain does not necessarily have to be straight.

2. Number the carbons in the chain so that the double bond would be between the carbons with the lowest designated number. This means that you have to decide whether to number beginning on the right end or left end of the chain. If it makes no difference to the double bond then shift attention to the branched groups.
3. Identify the various branching groups attached to this continuous chain of carbons by name
4. Name the branched groups in alphabetical order attaching (hyphenating) the carbon number it is attached to along the continuous chain of carbons to the front of the branch name. If more than one of the same kind of branched group is attached to the chain, identify the number carbon each group is attached to as a series of numbers separated by commas between each number then a hyphen and finally use a greek prefix attached to the branch name.
5. Attach a numerical prefix indicating the lowest carbon number the double bond is between onto the normal alkane name
6. Drop the "ane" ending and add the "ene" ending associated with the Alkene family

Here is an example:

To Identify the IUPAC name for the following:

$$
\begin{array}{c}
CH_2=CH\text{-}CH\text{-}CH\text{-}CH_3 \\
\quad\;\; | \;\;\; | \\
\quad\;\; CH_3Br
\end{array}
$$

1. Identify the longest continuous chain of carbons with the double bond between two of them

   Here we have five carbons

2. Number the carbon chain so the double bond is between the lowest numbered carbons

   Numbering from left to right would place the double bond between carbon #1 and carbon #2.

3. Identify and locate all branched groups, alphabetically, attaching a prefixed number equal to the carbon number the branch is attached to.

   Here we have a methyl group attached to the #3 carbon and a halogen attached to the #4 carbon. Halogens are named as branches using the following:

   - -F Fluoro
   - -Cl Chloro
   - -Br Bromo
   - -I Iodo

   So we would have 4-Bromo-3-methyl

1. Attach a numerical prefix (equal to the lowest carbon number in which the double bond is between) to the normal alkane name corresponding to the number of carbons that were in the continuous chain

   4-Bromo-3-methyl-1-Pentane

2. change the "ane" ending to "ene"

   4-Bromo-3-methyl-1-Pentene

Isomerism

Structural Isomerism

All the alkenes with 4 or more carbon atoms in them show *structural isomerism.* This means that there are two or more different structural formulae that you can draw for each molecular formula.

For example, with C_4H_8, it isn't too difficult to come up with these three structural isomers:

There is, however, another isomer. But-2-ene also exhibits geometric isomerism.

Geometric (cis-trans) Isomerism

The carbon-carbon double bond doesn't allow any rotation about it. That means that it is possible to have the CH_3 groups on either end of the molecule locked either on one side of the molecule or opposite each other.

These are called cis-but-2-ene (where the groups are on the same side) or trans-but-2-ene (where they are on opposite sides).

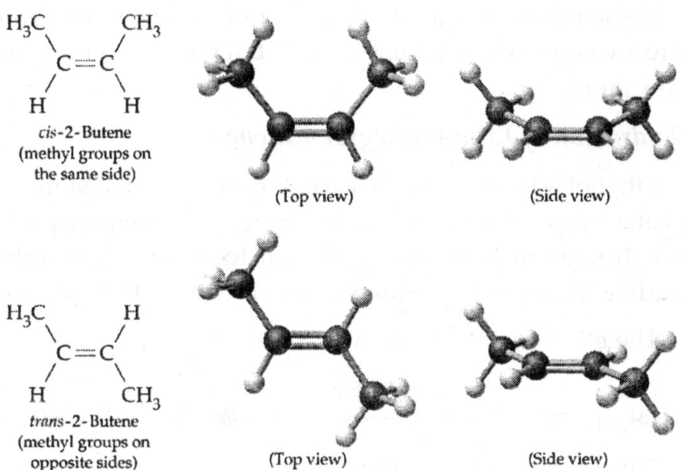

Fig. Geometrical isomers of butene

Preparation

Dehydration of Alcohols using Aluminium Oxide as Catalyst

The dehydration of ethanol to give ethene

This is a simple way of making gaseous alkenes like ethene. If ethanol vapour is passed over heated aluminium oxide powder, the ethanol is essentially cracked to give ethene and water vapour.

$$CH_3\text{-}CH_2\text{-}OH \xrightarrow{Al_2O_3} CH_2{=}CH_2 + H_2O$$

Dehydration of Alcohols using an Acid Catalyst

The acid catalysts normally used are either concentrated sulphuric acid or concentrated phosphoric(V) acid, H_3PO_4.

Concentrated sulphuric acid produces messy results. Not only is it an acid, but it is also a strong oxidizing agent. It oxidizes some of the alcohol to carbon dioxide and at the same time is reduced itself to sulphur dioxide. Both of these gases have to be removed from the alkene.

It also reacts with the alcohol to produce a mass of carbon. There are other side reactions as well, but these aren't required at this point.

Dehydration of Ethanol to give Ethene

Ethanol is heated with an excess of concentrated sulphuric acid at a temperature of 170°C. The gases produced are passed through sodium hydroxide solution to remove the carbon dioxide and sulphur dioxide produced from side reactions.

The ethene is collected over water.

$$CH_3\text{-}CH_2\text{-}OH \xrightarrow{\text{Conc } H_2SO_4} CH_2 = CH_2 + H_2O$$

The concentrated sulphuric acid is a catalyst.

Dehydration of Cyclohexanol to give Cyclohexene

This is a preparation commonly used at this level to illustrate the formation and purification of a liquid product. The fact that the carbon atoms happen to be joined in a ring makes no difference whatever to the chemistry of the reaction.

Cyclohexanol is heated with concentrated phosphoric(V) acid and the liquid cyclohexene distils off and can be collected and purified.

Phosphoric(V) acid tends to be used in place of sulphuric acid because it is safer and produces a less messy reaction.

$$\text{cyclohexanol } (C_6H_{11}OH) \xrightarrow{\text{conc } H_3PO_4} \text{cyclohexene } (C_6H_{10}) + H_2O$$

cyclohexanol — cyclohexene

Bonding in Ethene

Ethene, C_2H_4

The simple view of the bonding in ethene

At a simple level, you will have drawn ethene showing two bonds between the carbon atoms. Each line in this diagram represents one pair of shared electrons.

Ethene is actually much more interesting than this.

An orbital View of the Bonding in Ethene

Ethene is built from hydrogen atoms ($1s^1$) and carbon atoms ($1s^2 2s^2 2p_x^{\ 1} 2p_y^{\ 1}$).

The carbon atom doesn't have enough unpaired electrons to form the required number of bonds, so it needs to promote one of the $2s^2$ pair into the empty $2p_z$ orbital. This is exactly the same as happens whenever carbon forms bonds - whatever else it ends up joined to.

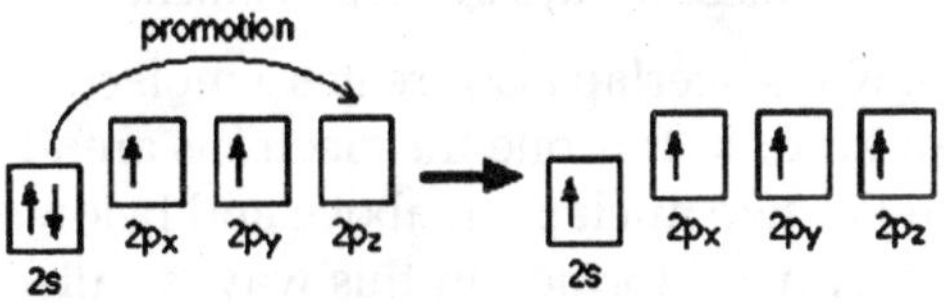

Now there's a difference, because each carbon is only joining to three other atoms rather than four - as in methane

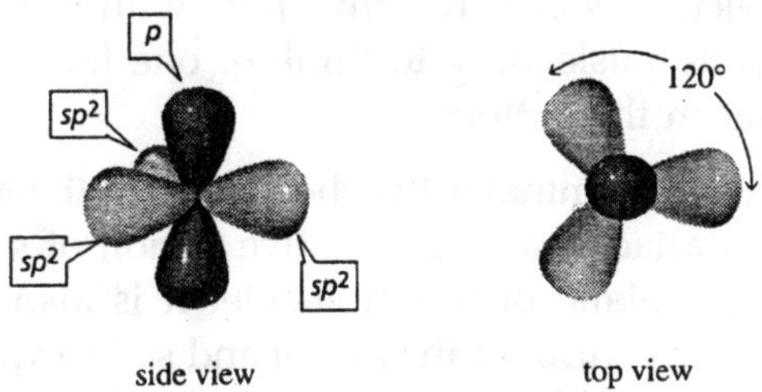

Fig. sp2 hybrid orbitals

or ethane. When the carbon atoms hybridise their outer orbitals before forming bonds, this time they only hybridise *three* of the orbitals rather than all four. They use the 2s electron and two of the 2p electrons, but leave the other 2p electron unchanged.

The new orbitals formed are called *sp^2 hybrids,* because they are made by an s orbital and two p orbitals reorganising themselves. sp^2 orbitals look rather like sp^3 orbitals that you have already come across in the bonding in methane, except that they are shorter and fatter. The three sp^2 hybrid orbitals arrange themselves as far apart as possible - which is at 120° to each other in a plane. The remaining p orbital is at right angles to them.

Notice that the p orbitals are so close that they are overlapping sideways.

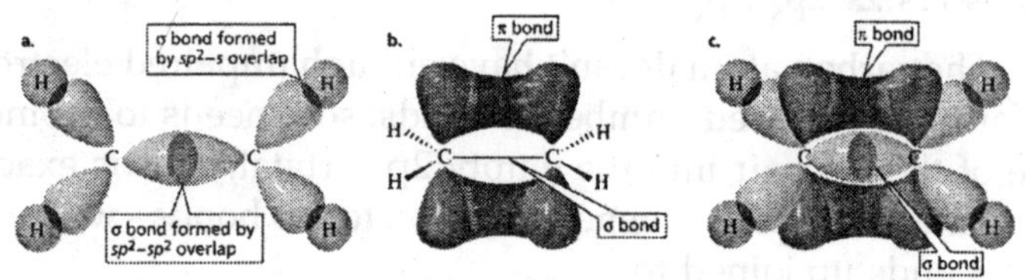

Fig. Sideways overlap in ethene

This sideways overlap also creates a molecular orbital, but of a different kind. In this one the electrons aren't held on the line between the two nuclei, but above and below the plane of the molecule. A bond formed in this way is called a *pi bond.*

Be clear about what a pi bond is. It is a region of space in which you can find the two electrons which make up the bond. Those two electrons can live anywhere within that space. It would be quite misleading to think of one living in the top and the other in the bottom.

The pi bond dominates the chemistry of ethene. It is very vulnerable to attack - a very negative region of space above and below the plane of the molecule. It is also somewhat distant from the control of the nuclei and so is a weaker bond than the sigma bond joining the two carbons.

All double bonds (whatever atoms they might be joining) will consist of a sigma bond and a pi bond.

Physical Properties of Alkenes

The boiling points and solubilities of alkenes are very similar to the alkanes. Both families have members that are non-polar. Alkenes are also referred to as "olefins". The word means fat dissolving since fats are relatively non-polar as alkenes are.

Boiling Points

The boiling point of each alkene is very similar to that of the alkane with the same number of carbon atoms. Ethene, propene and the various butenes are gases at room temperature. All the rest that you are likely to come across are liquids.

In each case, the alkene has a boiling point which is a small number of degrees lower than the corresponding alkane. The only attractions involved are Van der Waals dispersion forces, and these depend on the shape of the molecule and the number of electrons it contains. Each alkene has 2 fewer electrons than the alkane with the same number of carbons.

Solubility

Alkenes are virtually insoluble in water, but dissolve in organic solvents.

Reactions of Alkenes

Like any other hydrocarbons, alkenes burn in air or oxygen, but these reactions are unimportant. Alkenes are too valuable to waste in this way.

The important reactions all centre around the double bond. Typically, the pi bond breaks and the electrons from it are used to join the two carbon atoms to other things. Alkenes undergo addition reactions.

For example, using a general molecule X-Y . . .

$$H_2C{=}CH_2 + X{-}Y \longrightarrow H{-}\underset{X}{\overset{H}{C}}{-}\underset{Y}{\overset{H}{C}}{-}H$$

The rather exposed electrons in the pi bond are particularly open to attack by things which carry some degree of positive charge. These are called *electrophiles.*

Hydrogenation

Here we will discuss the reaction of the carbon-carbon double bond in alkenes with hydrogen in the presence of a metal catalyst. This is called hydrogenation. It includes the manufacture of margarine from animal or vegetable fats and oils.

Hydrogenation in the Lab

Hydrogenation of Ethene

Ethene reacts with hydrogen in the presence of a finely divided nickel catalyst at a temperature of about 150°C. Ethane is produced.

$$CH_2{=}CH_2 + H_2 \xrightarrow{Ni} CH_3\text{-}CH_3$$

This is a fairly pointless reaction because ethene is a far more useful compound than ethane! However, what is true of the reaction of the carbon-carbon double bond in ethene is equally true of it in much more complicated cases.

Animal and Vegetable Fats and Oils

These are similar molecules, differing in their melting points. If the compound is a solid at room temperature, you usually call it a fat. If it is a liquid, it is often described as an oil.

Their melting points are largely determined by the presence of carbon-carbon double bonds in the molecule. The higher the number of carbon-carbon double bonds, the lower the melting point.

If there aren't any carbon-carbon double bonds, the substance is said to be saturated. A typical saturated fat might have the structure:

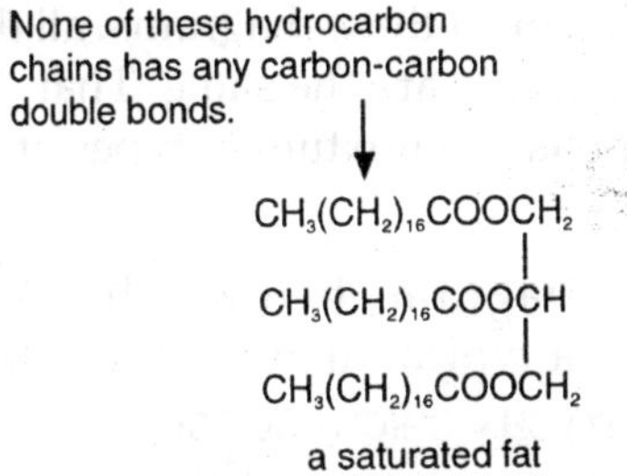

None of these hydrocarbon chains has any carbon-carbon double bonds.

$$
\begin{array}{r}
CH_3(CH_2)_{16}COOCH_2 \\
| \\
CH_3(CH_2)_{16}COOCH \\
| \\
CH_3(CH_2)_{16}COOCH_2
\end{array}
$$

a saturated fat

Molecules of this sort are usually solid at room temperature.

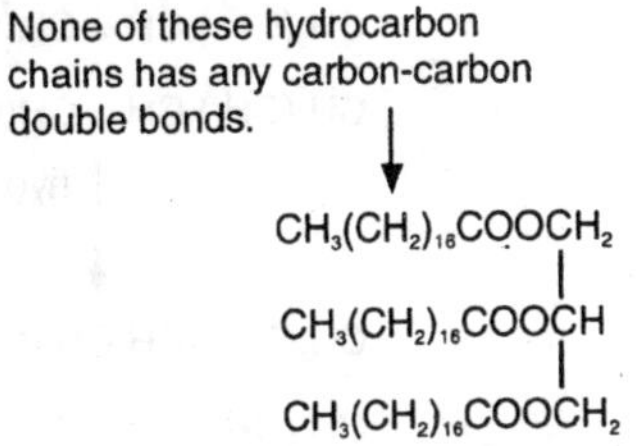

None of these hydrocarbon chains has any carbon-carbon double bonds.

$$
\begin{array}{r}
CH_3(CH_2)_{16}COOCH_2 \\
| \\
CH_3(CH_2)_{16}COOCH \\
| \\
CH_3(CH_2)_{16}COOCH_2
\end{array}
$$

a saturated fat

If there is only one carbon-carbon double bond in each of the hydrocarbon chains, it is called a *mono-unsaturated fat* (or mono-unsaturated oil, because it is likely to be a liquid at room temperature.)

A typical mono-unsaturated oil might be:

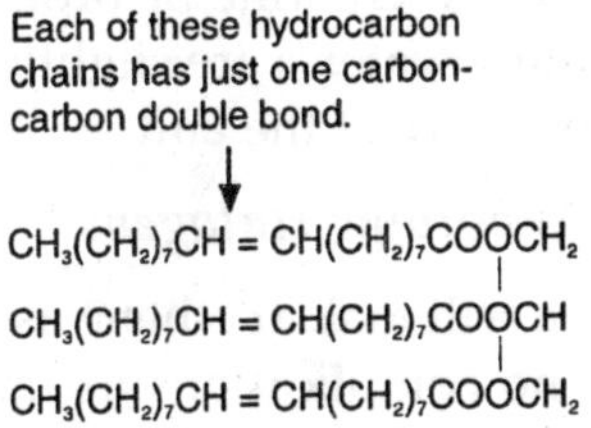

Each of these hydrocarbon chains has just one carbon-carbon double bond.

$$
\begin{array}{r}
CH_3(CH_2)_7CH = CH(CH_2)_7COOCH_2 \\
| \\
CH_3(CH_2)_7CH = CH(CH_2)_7COOCH \\
| \\
CH_3(CH_2)_7CH = CH(CH_2)_7COOCH_2
\end{array}
$$

a mono unsaturated oil

If there are two or more carbon-carbon double bonds in each chain, then it is said to be *polyunsaturated*.

For example:

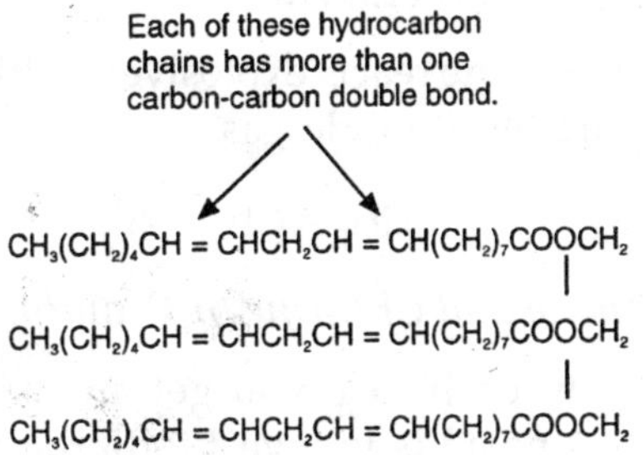

Each of these hydrocarbon chains has more than one carbon-carbon double bond.

$$
\begin{array}{r}
CH_3(CH_2)_4CH = CHCH_2CH = CH(CH_2)_7COOCH_2 \\
| \\
CH_3(CH_2)_4CH = CHCH_2CH = CH(CH_2)_7COOCH_2 \\
| \\
CH_3(CH_2)_4CH = CHCH_2CH = CH(CH_2)_7COOCH_2
\end{array}
$$

a polyunsaturated oil

For simplicity, in all these diagrams, all three hydrocarbon chains in each molecule are the same. That doesn't have to be the case - you can have a mixture of types of chain in the same molecule.

The flow diagram below shows the complete hydrogenation of a typical mono-unsaturated oil.

$CH_3(CH_2)_7CH = CH(CH_2)_7COOCH_2$

$CH_3(CH_2)_7CH = CH(CH_2)_7COOCH$ saturated oil

$CH_3(CH_2)_7CH = CH(CH_2)_7COOCH_2$

↓ hydrogen, heat, nickel catalyst

$CH_3(CH_2)_7CH_2CH_2(CH_2)_7COOCH_2$

$CH_3(CH_2)_7CH_2CH_2(CH_2)_7COOCH$ saturated fat

$CH_3(CH_2)_7CH_2CH_2CH_2)_7COOCH_2$

Halogenation

The reaction of the carbon-carbon double bond in alkenes such as ethene with halogens such as chlorine, bromine and iodine is called halogenation.

Reactions where the chlorine or bromine are in solution (for example, "bromine water") are slightly more complicated and are treated separately at the end.

Simple Reactions Involving Halogens

In each case, we will look at ethene as typical of all of the alkenes. There are no complications as far as the basic facts are concerned as the alkenes get bigger.

Ethene and Fluorine

Ethene reacts explosively with fluorine to give carbon and hydrogen fluoride gas.

$$CH_2 = CH_2 + 2F_2 \longrightarrow 2C + 4HF$$

Ethene and Chlorine or Bromine or Iodine

In each case you get an *addition reaction*. For example, bromine adds to give 1,2-dibromoethane.

$$CH_2{=}CH_2 + Br_2 \longrightarrow \underset{Br}{CH_2}\text{-}\underset{Br}{CH_2}$$

The reaction with bromine happens at room temperature. If you have a gaseous alkene like ethene, you can bubble it through either pure liquid bromine or a solution of bromine in an organic solvent like tetrachloromethane. The reddish-brown bromine is decolourized as it reacts with the alkene.

A liquid alkene (like cyclohexene) can be shaken with liquid bromine or its solution in tetrachloromethane.

Chlorine reacts faster than bromine, but the chemistry is similar. Iodine reacts much, much more slowly, but again the chemistry is similar. You are much more likely to meet the bromine case than either of these.

With Bromine Water

Using Bromine Water as a Test for Alkenes

If you shake an alkene with bromine water (or bubble a gaseous alkene through bromine water), the solution becomes colourless. Alkenes decolorize bromine water.

The Chemistry of the Test

This is complicated by the fact that the major product isn't 1,2-dibromoethane. The water also gets involved in the reaction, and most of the product is 2-bromoethanol.

$$CH_2{=}CH_2 + Br_2 + H_2O \longrightarrow \underset{Br}{CH_2}\text{-}CH_2OH + HBr$$

However, there will still be some 1,2-dibromoethane formed, so at this sort of level you can probably get away with quoting the simpler equation:

$$CH_2{=}CH_2 + Br_2 \longrightarrow \underset{Br}{CH_2}\text{-}\underset{Br}{CH_2}$$

With Hydrogen Halides

Here we look at the reaction of the carbon-carbon double bond in alkenes such as ethene with hydrogen halides such as hydrogen chloride and hydrogen bromide.

Symmetrical alkenes (like ethene or but-2-ene) are dealt with first. These are alkenes where identical groups are attached to each end of the carbon-carbon double bond. The extra problems associated with unsymmetrical ones like propene are covered in a separate section afterwards.

Addition to Symmetrical Alkenes

All alkenes undergo addition reactions with the hydrogen halides. A hydrogen atom joins to one of the carbon atoms originally in the double bond, and a halogen atom to the other.

For example, with ethene and hydrogen chloride, you get chloroethane:

$$CH_2 = CH_2 + HCl \longrightarrow CH_3\text{-}CH_2Cl$$

With but-2-ene you get 2-chlorobutane:

$$CH_3\text{-}CH = CH\text{-}CH_3 + HCl \longrightarrow CH_3\text{-}CH_2\text{-}\underset{\displaystyle Cl}{\underset{|}{CH}}\text{-}CH_3$$

What happens if you add the hydrogen to the carbon atom at the right-hand end of the double bond, and the chlorine to the left-hand end? You would still have the same product.

The chlorine would be on a carbon atom next to the end of the chain - you would simply have drawn the molecule flipped over in space.

That would be different of the alkene was unsymmetrical - that's why we have to look at them separately.

Conditions

The alkenes react with gaseous hydrogen halides at room temperature. If the alkene is also a gas, you can simply mix the gases. If the alkene is a liquid, you can bubble the hydrogen halide through the liquid.

Alkenes will also react with concentrated solutions of the gases in water. A solution of hydrogen chloride in water is, of course, hydrochloric acid. A solution of hydrogen bromide in water is hydrobromic acid - and so on.

There are, however, problems with this. The water will also get involved in the reaction and you end up with a mixture of products.

Reaction Rates

Variation of Rates when you Change the Halogen

Reaction rates increase in the order HF - HCl - HBr - HI. Hydrogen fluoride reacts much more slowly than the other three, and is normally ignored in talking about these reactions.

When the hydrogen halides react with alkenes, the hydrogen-halogen bond has to be broken. The bond strength falls as you go from HF to HI, and the hydrogen-fluorine bond is particularly strong. Because it is difficult to break the bond between the hydrogen and the fluorine, the addition of HF is bound to be slow.

Variation of Rates when you Change the Alkene

This applies to unsymmetrical alkenes as well as to symmetrical ones. For simplicity the examples given below are all symmetrical ones- but they don't have to be.

Reaction rates increase as the alkene gets more complicated - in the sense of the number of alkyl groups (such as methyl groups) attached to the carbon atoms at either end of the double bond.

For example:

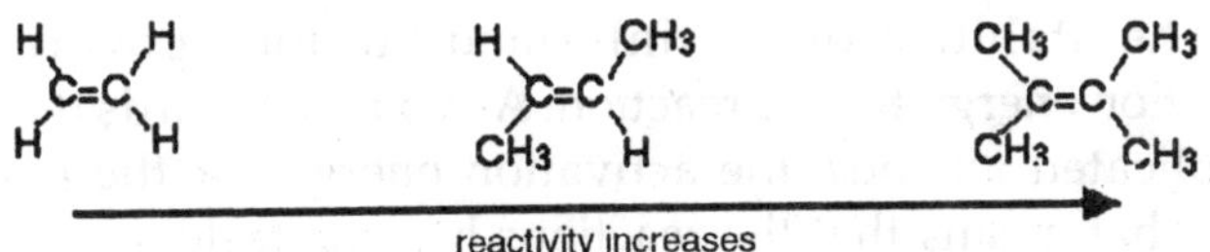

There are two ways of looking at the reasons for this - both of which need you to know about the mechanism for the reactions.

Alkenes react because the electrons in the pi bond attract things with any degree of positive charge. Anything which increases the electron density around the double bond will help this.

Alkyl groups have a tendency to "push" electrons away from themselves towards the double bond. The more alkyl groups you have, the more negative the area around the double bonds becomes.

The more negatively charged that region becomes, the more it will attract molecules like hydrogen chloride.

The more important reason, though, lies in the stability of the intermediate ion formed during the reaction. The three examples given above produce these carbocations (carbonium ions) at the half-way stage of the reaction:

$H_2C=CH_2$ ⟶ $CH_3\text{-}\overset{+}{C}H_2$

a primary carbocation

$(H)(CH_3)C=C(CH_3)(H)$ ⟶ $CH_3\text{-}CH_2\text{-}\overset{+}{C}H\text{-}CH_3$

a secondary carbocation

$(CH_3)_2C=C(CH_3)_2$ ⟶ $CH_3\text{-}CH(CH_3)\text{-}\overset{+}{C}(CH_3)\text{-}CH_3$

a tertiary carbocation

ions getting more energetically stable and so easier to form

The stability of the intermediate ions governs the activation energy for the reaction. As you go towards the more complicated alkenes, the activation energy for the reaction falls. That means that the reactions become faster.

Addition to Unsymmetrical Alkenes

In terms of reaction conditions and the factors affecting the rates of the reaction, there is no difference whatsoever between these alkenes and the symmetrical ones described above. The problem comes with the orientation of the addition - in other words, which way around the hydrogen and the halogen add across the double bond.

Orientation of Addition

If HCl adds to an unsymmetrical alkene like propene, there are two possible ways it could add. However, in practice, there is only one major product.

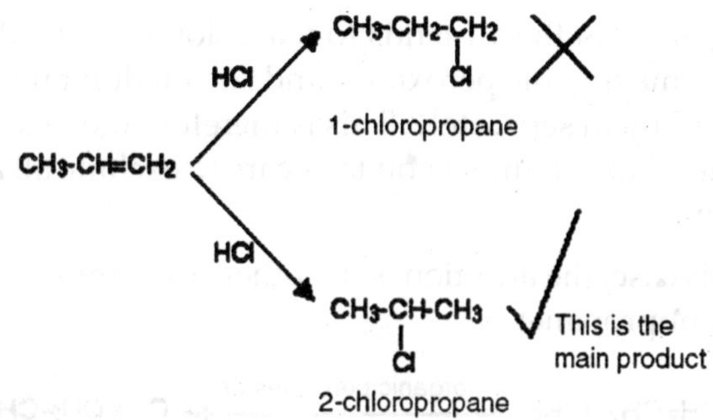

This is in line with *Markovnikov's Rule* which says:

When a compound HX is added to an unsymmetrical alkene, the hydrogen becomes attached to the carbon with the most hydrogens attached to it already.

In this case, the hydrogen becomes attached to the CH_2 group, because the CH_2 group has more hydrogens than the CH group.

Notice that only the hydrogens directly attached to the carbon atoms at either end of the double bond count. The ones in the CH_3 group are totally irreleVant.

A Special Problem with Hydrogen Bromide

Unlike the other hydrogen halides, hydrogen bromide can add to a carbon-carbon double bond *either* way around - depending on the conditions of the reaction.

If the Hydrogen Bromide and Alkene are Entirely Pure

In this case, the hydrogen bromide adds on according to Markovnikov's Rule. For example, with propene you would get 2-bromopropane.

$$CH_3\text{-}CH{=}CH_2 + HBr \longrightarrow CH_3\text{-}\underset{\displaystyle Br}{\underset{|}{CH}}\text{-}CH_3$$

2-bromopropane

That is exactly the same as the way the other hydrogen halides add.

If the Hydrogen Bromide and Alkene Contain Traces of Organic Peroxides

Oxygen from the air tends to react slowly with alkenes to produce some organic peroxides, and so you don't necessarily have to add them separately. This is therefore the reaction that you will tend to get unless you take care to exclude all air from the system.

In this case, the addition is the other way around, and you get 1-bromopropane:

$$CH_3\text{-}CH{=}CH_2 + HBr \xrightarrow[\text{Oxygen from the air}]{\text{organic peroxides or}} CH_3\text{-}CH_2\text{-}CH_3\text{-}Br$$

1-bromopropane

This is sometimes described as an *anti-Markovnikov addition* or as the *peroxide effect.*

Organic peroxides are excellent sources of free radicals. In the presence of these, the hydrogen bromide reacts with alkenes using a different (faster) mechanism. For various reasons, this doesn't happen with the other hydrogen halides.

This reaction can also happen in this way in the presence of ultra-violet light of the right wavelength to break the hydrogen-bromine bond into hydrogen and bromine free radical.

Reaction with Sulphuric Acid

Here we look at the reaction of the carbon-carbon double bond in alkenes such as ethene with concentrated sulphuric acid. It includes the conversion of the product into an alcohol.

Addition of Sulphuric Acid to Alkenes

Reaction with Ethene

Alkenes react with concentrated sulphuric acid in the cold to produce alkyl hydrogensulphates. Ethene reacts to give ethyl hydrogensulphate.

$$CH_2 = CH_2 + H_2SO_4 \longrightarrow CH_3CH_2OSO_2OH$$

The structure of the product molecule is sometimes written as $CH_3CH_2HSO_4$, but the version in the equation is better because it shows how all the atoms are linked up.

Reaction with Propene

This is typical of the reaction with unsymmetrical alkenes. An unsymmetrical alkene has different groups at either end of the carbon-carbon double bond.

If sulphuric acid adds to an unsymmetrical alkene like propene, there are two possible ways it could add. You could end up with one of two products depending on which carbon atom the hydrogen attaches itself to.

However, in practice, there is only one major product.

$$CH_3CH{=}CH_2 + H_2SO_4 \longrightarrow CH_3\underset{\displaystyle OSO_2OH}{\underset{|}{C}}HCH_3$$

This is in line with *Markovnikov's Rule* which says:

When a compound HX is added to an unsymmetrical alkene, the hydrogen becomes attached to the carbon with the most hydrogens attached to it already.

In this case, the hydrogen becomes attached to the CH_2 group, because the CH_2 group has more hydrogens than the CH group.

Notice that only the hydrogens directly attached to the carbon atoms at either end of the double bond count. The ones in the CH_3 group are totally irrelevant.

Using these reactions to make alcohols

Making ethanol

Ethene is passed into concentrated sulphuric acid to make ethyl hydrogensulphate (as above). The product is diluted with water and then distilled.

The water reacts with the ethyl hydrogensulphate to produce ethanol which distils off.

$$CH_3\text{-}CH_2\text{-}OSO_2OH + H_2O \xrightarrow{heat} CH_3\text{-}CH_2\text{-}OH + H_2SO_4$$

Making Propan-2-ol

More complicated alkyl hydrogensulphates react with water in exactly the same way. For example:

$$\underset{\displaystyle |\atop \displaystyle OSO_2OH}{CH_3\text{-}CH\text{-}CH_2} + H_2O \xrightarrow{heat} \underset{\displaystyle |\atop \displaystyle OH}{CH_3\text{-}CH\text{-}CH_2} + H_2SO_4$$

Notice that the position of the -OH group is determined by where the HSO_4 group was attached. You get propan-2-ol rather than propan-1-ol because of the way the sulphuric acid originally added across the double bond in propene.

Using these Reactions

These reactions were originally used as a way of manufacturing alcohols from alkenes in the petrochemical industry. These days, alcohols like ethanol or propan-2-ol tend to be manufactured by direct hydration of the alkene because it is cheaper and easier.

With Potassium Manganate(VII)

Lets look at the reaction of the carbon-carbon double bond in alkenes such as ethene with potassium manganate(VII) solution (potassium permanganate solution).

Oxidation of Alkenes

Alkenes react with potassium manganate(VII) solution in the cold. The colour change depends on whether the potassium manganate(VII) is used under acidic or alkaline conditions.

If the potassium manganate(VII) solution is acidified with dilute sulphuric acid, the purple solution becomes colourless.

If the potassium manganate(VII) solution is made slightly alkaline (often by adding sodium carbonate solution), the purple solution first becomes dark green and then produces a dark brown precipitate.

Chemistry of the Reaction

We'll look at the reaction with ethene. Other alkenes react in just the same way.

Manganate(VII) ions are a strong oxidising agent, and in the first instance oxidize ethene to ethane-1,2-diol (old name: ethylene glycol).

Looking at the equation purely from the point of view of the organic reaction:

$$CH_2{=}CH_2 + H_2O + [O] \longrightarrow \underset{OH}{CH_2}\text{-}\underset{OH}{CH_2}$$

oxygen from the oxidising agent (↑ [O])

The full equation depends on the conditions.

Under acidic conditions, the manganate(VII) ions are reduced to manganese(II) ions.

$$5CH_2{=}CH_2 + 2H_2O + 2MnO_4^- + 6H^+ \longrightarrow 5\underset{OH}{CH_2}\text{-}\underset{OH}{CH_2} + 2Mn^{2+}$$

Under alkaline conditions, the manganate(VII) ions are first reduced to green manganate(VI) ions .

$$CH_2{=}CH_2 + 2MnO_4^- + 2OH^- \longrightarrow \underset{OH}{CH_2}\text{-}\underset{OH}{CH_2} + 2MnO_4^{2-}$$

dark green solution (↑ MnO_4^{2-})

and then further to dark brown solid manganese(IV) oxide (manganese dioxide).

$$3CH_2{=}CH_2 + 2MnO_4^- + 4H_2O \longrightarrow 3\underset{OH}{CH_2}\text{-}\underset{OH}{CH_2} + 2MnO_2 + 2OH^-$$

dark brown precipitate (↑ MnO_2)

This last reaction is also the one you would get if the reaction was done under neutral conditions. You will notice that there are neither hydrogen ions nor hydroxide ions on the left-hand side of the equation.

Complications

The product, ethane-1,2-diol, is itself quite easily oxidised by manganate(VII) ions, and so the reaction won't stop at this point unless the potassium manganate(VII) solution is *very* dilute, very cold, and preferably not under acidic conditions.

That means that this reaction has little use as a way of preparing ethane-1,2-diol. Its only real use is in testing for carbon-carbon double bonds.

Using theRreaction to Test for Carbon-Carbon Double Bonds

If an organic compound reacts with dilute alkaline potassium manganate(VII) solution to give a green solution followed by a dark brown precipitate, then it may contain a carbon-carbon double bond. But equally it could be any one of a large number of other compounds all of which can be oxidised by manganate(VII) ions under alkaline conditions.

The situation with acidified potassium manganate(VII) solution is even worse because it has a tendency to break carbon-carbon bonds. It reacts destructively with a large number of organic compounds and is rarely used in organic chemistry.

You *could* use alkaline potassium manganate(VII) solution if, for example, all you had to do was to find out whether a hydrocarbon was an alkane or an alkene - in other words, if there was nothing else present which could be oxidised.

Direct Hydration of Alkenes

Here we will discuss the production of alcohols by the direct hydration of alkenes - adding water directly to the carbon-carbon double bond.

Manufacturing Ethanol

Ethanol is manufactured by reacting ethene with steam. The reaction is reversible.

$$CH_2=CH_{2(g)} + H_2O_{(g)} \rightleftharpoons CH_3CH_2OH_{(g)}$$

Only 5% of the ethene is converted into ethanol at each pass through the reactor. By removing the ethanol from the equilibrium mixture and recycling the ethene, it is possible to achieve an overall 95% conversion.

Manufacturing other Alcohols

If you start from an unsymmetrical alkene like propene, you have to be careful to think about which way around the water adds across the carbon-carbon double bond.

Markovnikov's Rule says that when you add a molecule HX across a carbon-carbon double bond, the hydrogen joins to the carbon atom which already has the more hydrogen atoms attached to it.

Thinking of water as H-OH, the hydrogen will add to the carbon with the more hydrogens already attached. That means that in the propene case, you will get propan-2-ol rather than propan-1-ol.

$$CH_3\text{-}CH{=}CH_2 + H_2O \rightleftharpoons CH_3\text{-}\underset{\displaystyle OH}{\underset{|}{CH}}\text{-}CH_3$$

propan-2-ol

The conditions used during manufacture vary from alcohol to alcohol.

Polymerization of Alkenes

Under this heading we are going to look at the polymerization of alkenes to produce polymers like poly(ethene) (usually known as polythene, and sometimes as polyethylene), poly(propene) (old name: polypropylene), PVC and PTFE. It also looks briefly at how the structure of the polymers affects their properties and uses.

Poly(ethene) (polythene or polyethylene)

Low Density Poly(ethene): LDPE

Manufacture

This is an example of *addition polymerization.*

An addition reaction is one in which two or more molecules join together to give a single product. During the polymerization of ethene, thousands of ethene molecules join together to make poly(ethene) - commonly called polythene.

$$nCH_2 = CH_2 \longrightarrow [-CH_2-CH_2-]n$$

The number of molecules joining up is very variable, but is in the region of 2000 to 20000.

Conditions

Temperature: about 200°C

Pressure: about 2000 atmospheres

Initiator: a small amount of oxygen as an impurity

Properties and uses

Low density poly(ethene) has quite a lot of branching along the hydrocarbon chains, and this prevents the chains from lying tidily close to each other. Those regions of the poly(ethene) where the chains lie close to each other and are regularly packed are said to be *crystalline*. Where the chains are a random jumble, it is said to be *amorphous*. Low density poly(ethene) has a significant proportion of amorphous regions.

One chain is held to its neighbours in the structure by Van der Waals dispersion forces. Those attractions will be greater if the chains are close to each other. The amorphous regions where the chains are inefficiently packed lower the effectiveness of the Van der Waals attractions and so lower the melting point and strength of the polymer. They also lower the density of the polymer (hence: "low density poly(ethene)").Low density poly(ethene) is used for familiar things like plastic carrier bags and other similar low strength and flexible sheet materials.

High density poly(ethene): HDPE

Manufacture

This is made under quite different conditions from low density poly(ethene).

Conditions

Temperature: about 60°C

Pressure: low - a few atmospheres

Catalyst: Ziegler-Natta catalysts or other metal compounds

Ziegler-Natta catalysts are mixtures of titanium compounds like titanium(III) chloride, $TiCl_3$, or titanium(IV) chloride, $TiCl_4$, and compounds of aluminium like aluminium triethyl, $Al(C_2H_5)_3$. There are all sorts of other catalysts constantly being developed.

These catalysts work by totally different mechanisms from the high pressure process used to make low density poly(ethene). The chains grow in a much more controlled - much less random - way.

Properties and Uses

High density poly(ethene) has very little branching along the hydrocarbon chains - the crystallinity is 95% or better. This better packing means that Van der Waals attractions between the chains are greater and so the plastic is stronger and has a higher melting point. Its density is also higher because of the better packing and smaller amount of wasted space in the structure.

Poly(chloroethene) (polyvinyl chloride): PVC

Poly(chloroethene) is commonly known by the initials of its old name, PVC.

Poly(chloroethene) is made by polymerising chloroethene, CH_2=CHCl.

The equation is usually written:

$$n\,\mathrm{C(Cl)(H)}{=}\mathrm{C(H)(H)} \longrightarrow \left(\mathrm{-C(Cl)(H)-C(H)(H)-}\right)_n$$

It doesn't matter which carbon you attach the chlorine to in the original molecule. Just be consistent on both sides of the equation.

Properties and Uses

You normally expect amorphous polymers to be more flexible than crystalline ones because the forces of attraction between the chains tend to be weaker. However, pure poly(chloroethene) tends to be rather hard and rigid.

This is because of the presence of additional dipole-dipole interactions due to the polarity of the carbon-chlorine bonds. Chlorine is more electronegative than carbon, and so attracts the electrons in the bond towards itself. That makes the chlorine atoms slightly negative and the carbons slightly positive.

These permanent dipoles add to the attractions due to the temporary dipoles which produce the dispersion forces.

Plasticisers are added to the poly(chloroethene) to reduce the effectiveness of these attractions and make the plastic more flexible. The more plasticiser you add, the more flexible it becomes.

Poly(chloroethene) is used to make a wide range of things including guttering, plastic windows, electrical cable insulation, sheet materials for flooring and other uses, footwear, clothing, and so on and so on.

Poly(tetrafluoroethene): PTFE

You may have come across this under the brand names of Teflon or Fluon.

Structurally, PTFE is just like poly(ethene) except that each hydrogen in the structure is replaced by a fluorine atom.

The PTFE chains tend to pack well and PTFE is fairly crystalline. Because of the fluorine atoms, the chains also contain more electrons (for an equal length) than a corresponding poly(ethene) chain. Taken together (the good packing and the extra electrons) that means that the Van der Waals dispersion forces will be stronger than in even high density poly(ethene).

Epoxyethane (Ethylene Oxide)

The Manufacture of Epoxyethane

Conditions

Temperature:	about 250 - 300°C
Pressure:	about 15 atmospheres
Catalyst:	silver

$$2\,CH_2{=}CH_2 + O_2 \longrightarrow 2\,\underset{\diagdown O \diagup}{CH_2{-}CH_2}$$

Chapter 7

Alkyne

Introduction

have at least one triple co-valent bond between two adjacent carbons. The general Formula for the Alkyne is:

C_nH_{2n-2}

Where n is the number of carbon atoms.

$HC \equiv CH$	$CH_3C \equiv CH$	$\overset{1}{C}H_3\overset{2}{C} \equiv \overset{3}{C}\overset{4}{C}H_3$	$\overset{4}{C}H_3\overset{3}{C}H_2\overset{2}{C} \equiv \overset{1}{C}H$
Ethyne (Acetylene)	Propyne	2-Butyne	1-Butyne

Fig. Some common alkynes

IUPAC Nomenclature of Alkynes

1. Determine the longest continuous chain of carbons that have the triple bond between two of its carbons. By "longest continuous chain" is meant to be able to trace through the carbons without raising the tracer (or finger) off the surface. The chain does not necessarily have to be straight.

2. Number the carbons in the chain so that the triple bond would be between the carbons with the lowest designated number. This means that you have to decide whether to number beginning on the right end or left end of the chain. If it makes no difference to the triple bond then shift attention to the branched groups.

3. Identify the various branching groups attached to this continuous chain of carbons by name
4. Name the branched groups in alphabetical order attaching (hyphenating) the carbon number it is attached to along the continuous chain of carbons to the front of the branch name. If more than one of the same kind of branched group is attached to the chain, identify the number carbon each group is attached to as a series of numbers separated by commas between each number then a hyphen and finally use a greek prefix attached to the branch name.
5. Attach a numerical prefix indicating the lowest carbon number the triple bond is between onto the normal alkane name
6. Drop the "ane" ending and add the "yne" ending associated with the Alkene family

Preparation

1. Dehydrogenation of an alkane or alkene(Fig a)
2. Dehalogenation of a tetrahaloalkane (Fig b)
3. Dehydrohalogenation of a dihaloalkane (Fig c)

(a) $CH_3\text{-}CH_2\text{-}CH_3 \xrightarrow{Pt} CH_3\text{-}CH{=}CH_2 \xrightarrow{Pt} CH_3\text{-}C{\equiv}CH$

(b) $CH_3\text{-}C(Cl)_2\text{-}CHCl_2 + 2\,Zn \xrightarrow{CH_3COOH} CH_3\text{-}C{\equiv}CH + 2ZnCl_2$

(c) $CH_3\text{-}CH(Cl)\text{-}CH_2(Cl) + 2\,KOH \xrightarrow{Heat} CH_3\text{-}C{\equiv}CH + 2KCl + 2H_2O$

Fig. (a) (b) (c)

Bonding in Ethyne (Acetylene)

Ethyne, C_2H_2

The Simple View of the Bonding in Ethyne

Ethyne has a triple bond between the two carbon atoms. In the diagram each line represents one pair of shared electrons.

$$H-C\equiv C-H$$

Electronic Configuration of the Bonding in Ethyne

Ethyne is built from hydrogen atoms ($1s^1$) and carbon atoms ($1s^2 2s^2 2p_x{}^1 2p_y{}^1$).

The carbon atom doesn't have enough unpaired electrons to form four bonds (1 to the hydrogen and three to the other carbon), so it needs to promote one of the $2s^2$ pair into the empty $2p_z$ orbital. This is exactly the same as happens whenever carbon forms bonds - whatever else it ends up joined to.

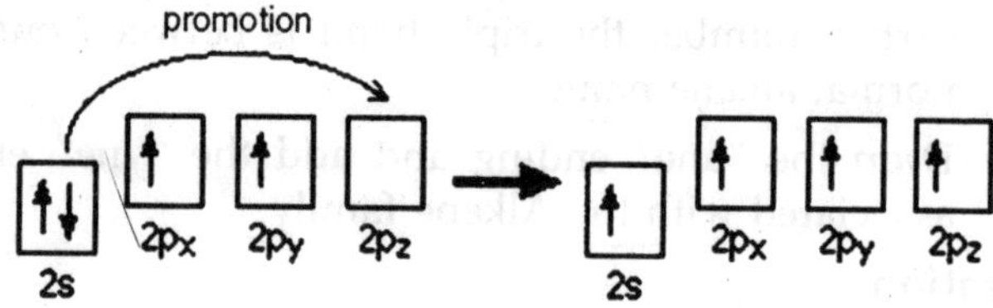

Each carbon is only joining to two other atoms rather than four (as in methane or ethane) or three (as in ethene) and so when the carbon atoms hybridise their outer orbitals before forming bonds, this time they only hybridise *two* of the orbitals.

They use the 2s electron and one of the 2p electrons, but leave the other 2p electrons unchanged. The new hybrid orbitals formed are called sp^1 hybrids (sometimes just sp hybrids), because they are made by an s orbital and a single p orbital reorganising themselves.

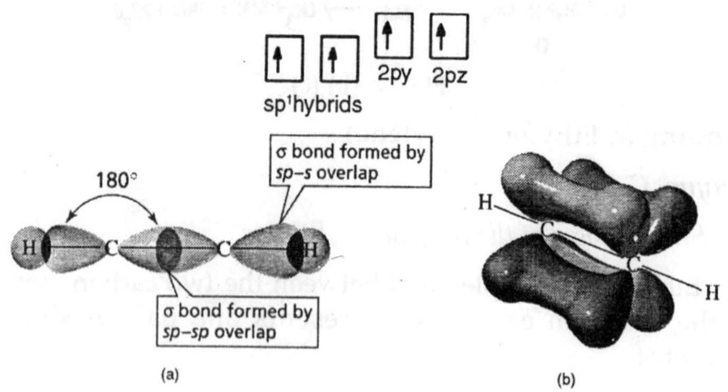

(a) (b)

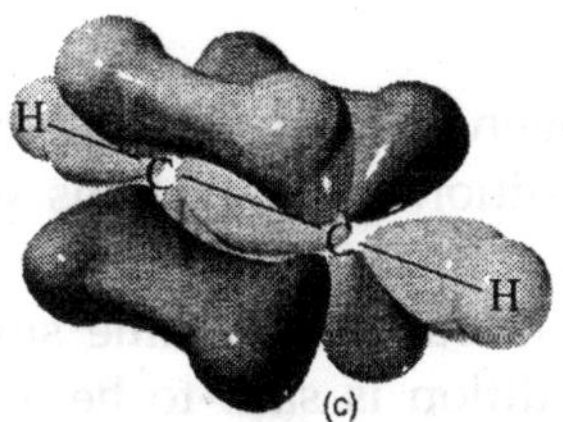

Fig. Bonding in ethyne

Sideways overlap between the two sets of p orbitals produces two pi bonds - each similar to the pi bond found in, say, ethene. These pi bonds are at 90° to each other - one above and below the molecule, and the other in front of and behind the molecule.

Physical Properties

As with hydrocarbons in general, alkynes are non-polar and are insoluble in water but soluble in non-polar organic solvents.

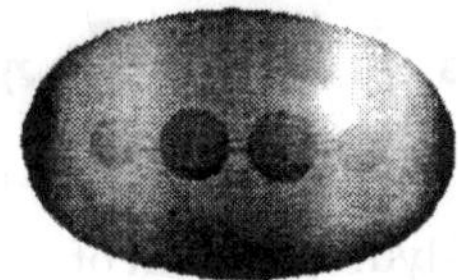

Reactivity

- The π bonds are a region of high electron density (red) so alkynes are typically nucleophiles.
- Alkynes typically undergo addition reactions in which one or both of the σ-bonds are converted to new Ã ?bonds.
- Terminal alkynes, R-Ca≡C-H, are quite acidic (indicated by blue) for hydrocarbons, pKa = 26
- Deprotonation of a terminal acetylene gives an acetylide ion.
- The acetylide ion is a good nucleophile and can be alkylated to give higher alkynes.

Reactions

1. Syn Hydrogenation of An Alkyne: There are two kinds of addition type reactions where a Pi bond is broken and atoms are added to the molecule. If the atoms are added on the same side of the molecule then the addition is said to be a "syn" addition. If the added atoms are added on opposite sides of the molecule then the addition is said to be an "anti" addition. Hydrogen atoms can be added to an alkyne on a one mole to one mole ratio to get an alkene where the Hydrogen atoms have been added on the same side of the molecule. This is called syn Hydrogenation.(Fig 1-a)
2. Anti Hydrogenation of An Alkyne (Fig 1-b)

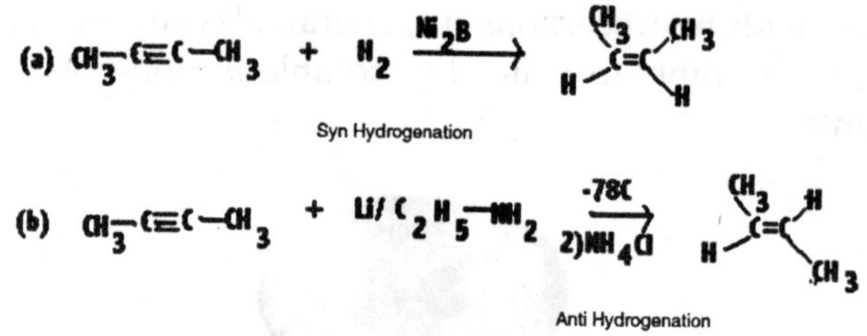

(a) $CH_3-C\equiv C-CH_3 + H_2 \xrightarrow{Ni_2B}$ (CH$_3$)(H)C=C(CH$_3$)(H)

Syn Hydrogenation

(b) $CH_3-C\equiv C-CH_3 + Li/C_2H_5-NH_2 \xrightarrow[2) NH_4Cl]{-78C}$ (CH$_3$)(H)C=C(H)(CH$_3$)

Anti Hydrogenation

Fig. 1-Hydrogenation of Alkyne

3. Complete Hydrogenation of Alkynes (Fig 2-a)
4. Halogenation of Alkynes (Fig 2-b)

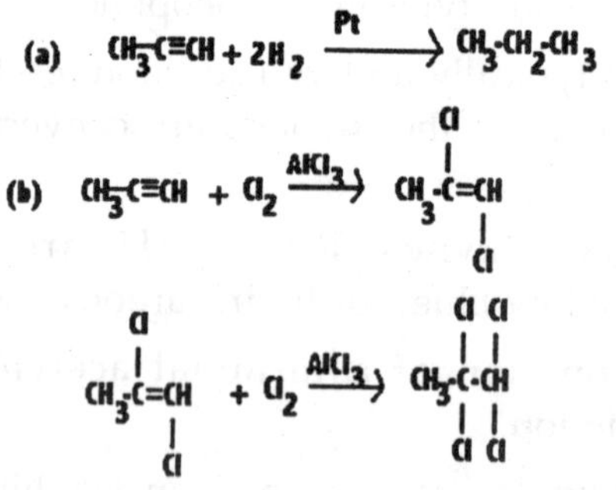

(a) $CH_3-C\equiv CH + 2H_2 \xrightarrow{Pt} CH_3-CH_2-CH_3$

(b) $CH_3-C\equiv CH + Cl_2 \xrightarrow{AlCl_3} CH_3-C(Cl)=CH(Cl)$

$CH_3-C(Cl)=CH(Cl) + Cl_2 \xrightarrow{AlCl_3} CH_3-C(Cl)_2-CH(Cl)_2$

Fig. 2-Complete Hydrogenation and Halogenation

5. Hydrohalogenation of Alkynes (Fig 4-a)

6. Oxidative Cleavage of Alkynes (Fig 4-b)

(a) $$CH_3\text{-}C{\equiv}CH + HX \longrightarrow CH_3\text{-}\underset{\underset{Cl}{|}}{C}{=}CH_2$$

$$CH_3\text{-}\underset{\underset{Cl}{|}}{C}{=}CH_2 + HX \longrightarrow CH_3\text{-}\overset{\overset{Cl}{|}}{\underset{\underset{Cl}{|}}{C}}\text{-}CH_3$$

(b) $$CH_3\text{-}C{\equiv}CH + O_3 \xrightarrow{2) H_2O} CH_3\text{-}\overset{\overset{H}{|}}{C}{=}O + H\text{-}\overset{\overset{H}{|}}{C}{=}O$$

or

$$CH_3\text{-}C{\equiv}CH + KMnO_4/OH^- \xrightarrow{2) H^+} CH_3\text{-}\overset{\overset{OH}{|}}{C}{=}O + H\text{-}\overset{\overset{OH}{|}}{C}{=}O$$

Fig. 4-Hydrohalogenation and Oxidative Cleavage of Alkyne

Addition Reactions

A carbon-carbon triple bond may be located at any unbranched site within a carbon chain or at the end of a chain, in which case it is called terminal. Because of its linear configuration (the bond angle of a sp-hybridized carbon is 180°), a ten-membered carbon ring is the smallest that can accomodate this function without excessive strain. Since the most common chemical transformation of a carbon-carbon double bond is an addition reaction, we might expect the same to be true for carbon-carbon triple bonds. Indeed, most of the alkene addition reactions discussed earlier also take place with alkynes, and with similar regio- and stereoselectivity.

Catalytic Hydrogenation

The catalytic addition of hydrogen to 2-butyne not only serves as an example of such an addition reaction, but also provides heat of reaction data that reflect the relative thermodynamic stabilities of these hydrocarbons, as shown in the diagram to the right. From the heats of hydrogenation, shown in blue in units of kcal/mole, it would appear that alkynes are thermodynamically less stable than alkenes to a greater degree than alkenes are less stable than alkanes. The standard bond energies for carbon-carbon bonds confirm this conclusion. Thus, a double bond is stronger than a single bond, but not twice as strong. The difference (63 kcal/mole) may be regarded as the strength of the À-bond component. Similarly, a triple bond is stronger than a double bond, but not 50% stronger. Here the difference (54 kcal/mole) may be taken as the strength of the second À-bond. The 9 kcal/mole weakening of this second À-bond is reflected in the heat of hydrogenation numbers (36.7 - 28.3 = 8.4).

Since alkynes are thermodynamically less stable than alkenes, we might expect addition reactions of the former to be more exothermic and relatively faster than equivalent reactions of the latter. In the case of catalytic hydrogenation, the usual Pt and Pd hydrogenation catalysts are so effective in promoting addition of hydrogen to both double and triple carbon-carbon bonds that the alkene intermediate formed by hydrogen addition to an alkyne cannot be isolated. A less efficient catalyst, Lindlar's catalyst, prepared by deactivating (or poisoning) a conventional palladium catalyst by treating it with lead acetate and quinoline, permits alkynes to be converted to alkenes without further reduction to an alkane. The addition of hydrogen is stereoselectively syn (e.g. 2-butyne gives cis-2-butene). A complementary stereoselective reduction in the anti mode may be accomplished by a solution of sodium in liquid ammonia. This reaction will be discussed later in this section.

R-C ≡ C-R + H_2 and Lindlar catalyst ⟶ *cis* R-CH=CH-R

R-CαC-R + 2 Na in NH_3 (liq) ⟶ *trans* R-CH=CH-R +2 $NaNH_2$

Alkenes and alkynes show a curious difference in behavior toward catalytic hydrogenation. Independent studies of hydrogenation rates for each class indicate that alkenes react more rapidly than alkynes. However, careful hydrogenation of an alkyne proceeds exclusively to the alkene until the former is consumed, at which point the product alkene is very rapidly hydrogenated to an alkane. This behavior is nicely explained by differences in the stages of the hydrogenation reaction. Before hydrogen can add to a multiple bond the alkene or alkyne must be adsorbed on the catalyst surface. Alkynes adsorb more strongly than alkenes, and preferentially occupy reactive sites on the catalyst. Subsequent transfer of hydrogen to the adsorbed alkyne proceeds slowly, relative to the corresponding hydrogen transfer to an adsorbed alkene molecule. Consequently, reduction of triple bonds occurs selectively at a moderate rate, followed by rapid addition of hydrogen to the alkene product. The Lindlar catalyst permits adsorbtion and reduction of alkynes, but does not adsorb alkenes sufficiently to allow their reduction.

Electrophilic Addition

When the addition reactions of electrophilic reagents, such as strong Brønsted acids and halogens, to alkynes are studied we find a curious paradox. The reactions are even more exothermic than the additions to alkenes, and yet the rate of addition to alkynes is slower by a factor of 100 to 1000 than addition to equivalently substituted alkenes. The reaction of one equivalent of bromine with 1-penten-4-yne, for example, gave 4,5-dibromo-1-pentyne as the chief product.

$$HC \equiv C\text{-}CH_2\text{-}CH{=}CH_2 + Br_2 \longrightarrow HCaHC\text{-}CH_2\text{-}CHBrCH_2Br$$

Although these electrophilic additions to alkynes are sluggish, they do take place and generally display Markovnikov Rule regioselectivity and anti-stereoselectivity. One problem, of course, is that the products of these additions are themselves substituted alkenes and can therefore undergo further addition. Because of their high electronegativity, halogen substituents on a double bond act to reduce its nucleophilicity, and thereby decrease the rate of electrophilic addition reactions. Consequently, there is a delicate balance as to whether the product of an initial addition to an alkyne will suffer further addition to a saturated product. Although the initial alkene products can often be isolated and identified, they are commonly present in mixtures of products and may not be obtained in high yield. The following reactions illustrate many of these features. In the last example, 1,2-diodoethene does not suffer further addition in as much as vicinal-diiodoalkanes are relatively unstable.

Why are the reactions of alkynes with electrophilic reagents more sluggish than the corresponding reactions of alkenes? After all, addition reactions to alkynes are generally more exothermic than additions to alkenes, and there would seem to be a higher À-electron density about the triple bond (two À-bonds versus one). Two factors are significant in explaining this apparent paradox. First, although there are more À-electrons associated with the triple bond, the sp-hybridized carbons exert a strong attraction for these À-electrons, which are consequently bound more tightly to the functional group than are the À-electrons of a double bond. This is seen in the ionization potentials of ethylene and acetylene.

Ethyne $HC \equiv CH$ + Energy $\longrightarrow$ $[HC \equiv CH] \bullet^{(+)} + e^{(-)}$
$\otimes H = +264$ kcal/mole

Ethylene $H_2C{=}CH_2$ + Energy $\longrightarrow$ $[H_2C{=}CH_2] \bullet^{(+)} + e^{(-)}$
$\Delta H = +244$ kcal/mole

Ethane $H_3C{-}CH_3$ + Energy $\longrightarrow$ $[H_3C{-}CH_3] \bullet^{(+)} + e^{(-)}$
$\Delta H = +296$ kcal/mole

As defined by the preceding equations, an ionization potential is the minimum energy required to remove an electron from a molecule of a compound. Since pi-electrons are less tightly held than sigma-electrons, we expect the ionization potentials of ethylene and acetylene to be lower than that of ethane, as is the case. Gas-phase proton affinities show the same order, with ethylene being more basic than acetylene, and ethane being less basic than either. Since the initial interaction between an electrophile and an alkene or alkyne is the formation of a pi-complex, in which the electrophile accepts electrons from and becomes weakly bonded to the multiple bond, the relatively slower reactions of alkynes becomes understandable.

The second factor is the stability of the carbocation intermediate generated by sigma-bonding of a proton or other electrophile to one of the triple bond carbon atoms. This intermediate has its positive charge localized on an unsaturated carbon, and such vinyl cations are less stable than their saturated analogs. Indeed, we can modify our earlier ordering of carbocation stability to include these vinyl cations.

Carbocation Stability

$CH_3^{(+)}$	$HRCH{=}CH^{(+)}$	$<RCH_2^{(+)}$	$HRCH{=}CR^{(+)}$	$<R_2CH^{(+)}H \approx$	$CH_2{=}CH\text{-}CH_2^{(+)}$	$<C_6H_5CH_2^{(+)}H \approx$	$R_3C^{(+)}$
Methyl	1°-Vinyl	1°	2°-Vinyl	2°	1°-Allyl	1°-Benzyl	3°

The activation energy for the generation of such an intermediate would be higher than that for a lower energy intermediate.

Despite these impediments, electrophilic additions to alkynes have emerged as exceptionally useful synthetic transforms. For example, addition of HCl, acetic acid and hydrocyanic acid to acetylene give respectively the useful monomers vinyl chloride, vinyl acetate and acrylonitrile, as shown in the following equations. Note that in these and many other similar reactions transition metals, such as copper and mercury salts, are effective catalysts.

$$HC \equiv CH + HCl \longrightarrow H_2C{=}CHCl \text{ \textit{Vinyl Chloride}}$$

$$HC \equiv CH + CH_3CO_2H \longrightarrow H_2C{=}CHOCOCH_3 \text{ \textit{Vinyl Acetate}}$$

$$HC \equiv HCH + HCN \longrightarrow H_2C{=}CHCN \text{ \textit{Acrylonitrile}}$$

Oxidations

Reactions of alkynes with oxidizing agents such as potassium permanganate and ozone usually result in cleavage of the triple-bond to give carboxylic acid products. A general equation for this kind of transformation follows. The symbol [O] is often used in a general way to denote an oxidation.

$$RC \equiv HCR' + [O] \longrightarrow RCO_2H + R'CO_2H$$

Nucleophilic Addition Reactions and Reduction

The sp-hybrid carbon atoms of the triple-bond render alkynes more electrophilic than similarly substituted alkenes. As a result, alkynes sometimes undergo addition reactions initiated by bonding to a nucleophile. This mode of reaction, illustrated below, is generally not displayed by alkenes, unless the double-bond is activated by electronegative substituents, *e.g.* $F_2C{=}CF_2$, or by conjugation with an electron withdrawing group.

$$HC \equiv CH + KOC_2H_5 \text{ in } C_2H_5OH \text{ at } 150°C \longrightarrow H_2C{=}CH\text{-}OC_2H_5$$

$$HC \equiv CH + HCN + NaCN \text{ (catalytic)} \longrightarrow H_2C{=}CH\text{-}CN$$

The smallest and most reactive nucleophilic species is probably an electron. Electron addition to a functional group is by definition a reduction, and we noted earlier that alkynes are reduced by solutions of sodium in liquid ammonia to trans-alkenes. To understand how this reduction occurs we first need to identify two distinct reactions of sodium with liquid ammonia (boiling point -78 °C). In the first, sodium dissolves in the pure liquid to give a deep blue solution consisting of very mobile and loosely bound electrons together with solvated sodium cations (first equation below). For practical purposes, we can consider such solutions to be a source of "free electrons" which may be used as powerful reducing agents. In the second case, ferric salts catalyze the reaction of sodium with ammonia, liberating hydrogen and forming the colorless salt sodium amide (second equation). This is analogous to the reaction of sodium with water to give sodium hydroxide, but since ammonia is 10^{18} times weaker an acid than water, the reaction is less violent. The usefulness of this reaction is that sodium amide, $NaNH_2$, is an exceedingly strong base (18 powers of ten stronger than sodium hydroxide), which may be used to convert very weak acids into their conjugate bases.

$$Na + NH_3 \text{ (liquid, –78 °C)} \longrightarrow Na^{(+)} + e^{(-)} \text{ (a blue solution)}$$

$$Na + NH_3 \text{ (liquid, –78 °C)} + Fe \longrightarrow H_2 + NaNH_2 \text{ (a colorless solution)}$$

Returning to the reducing capability of the blue electron solutions, we can write a plausible mechanism for the reduction of alkynes to trans-alkenes, as shown below. Isolated carbon double-bonds are not reduced by sodium in liquid ammonia, confirming the electronegativity difference between sp and sp^2 hybridized carbons.

Acidity of Terminal Alkynes

Alkanes are undoubtedly the weakest Brønsted acids commonly encountered in organic chemistry. It is difficult to measure such weak acids, but estimates put the pK_a of ethane at about 48. Hybridizing the carbon so as to increase the s-character of the C-H increases the acidity, with the greatest change occurring for the sp-C-H groups found in terminal alkynes. Thus, the pK_a of ethene is estimated at 44, and the pK_a of ethyne (acetylene) is found to be 25, making it 10^{23} times stronger an acid than ethane. This increase in acidity permits the isolation of insoluble silver and copper salts of such compounds.

$$RC \equiv C\text{-}H + Ag(NH_3)_2^{(+)} \xrightarrow{\text{(in } NH_4OH)} RC \equiv C\text{-}Ag \text{ (insoluble)} + NH_3 + NH_4^{(+)}$$

Despite the dramatic increase in acidity of terminal alkynes relative to other hydrocarbons, they are still very weak acids, especially when compared with water, which is roughly a billion times more acidic. If we wish to prepare nucleophilic salts of terminal alkynes for use in synthesis, it will therefore be necessary to use a much stronger base than hydroxide (or ethoxide) anion. Such a base is sodium amide ($NaNH_2$), discussed above, and its reactions with terminal alkynes may be conducted in liquid ammonia or ether as solvents. The products of this acid-base reaction are ammonia and a sodium acetylide salt. Because the acetylide anion is a powerful nucleophile it may displace halide ions from 1¡-alkyl halides to give a more highly substituted alkyne as a product (S_N2 reaction). This synthesis application is described in the following equations. The first two equations show how acetylene can be converted to propyne; the last two equations present a synthesis of 2-pentyne from propyne.

$$H\text{-}C\equiv C\text{-}H + NaNH_2 \text{ (in ammonia or ether)}$$

$$\longrightarrow H\text{-}C\equiv C\text{-}Na \text{ (sodium acetylide)} + NH_3$$

$$H\text{-}C\equiv C\text{-}Na + CH_3\text{-}I \longrightarrow H\text{-}C\equiv C\text{-}CH_3 + NaI$$

$$CH_3\text{-}C\equiv C\text{-}H + NaNH_2 \text{ (in ammonia or ether)}$$

$$\longrightarrow CH_3\text{-}C\equiv C\text{-}Na \text{ (sodium propynylide)} + NH_3$$

$$CH_3\text{-}C\equiv C\text{-}Na + C_2H_5\text{-}Br \longrightarrow CH_3\text{-}C\equiv C\text{-}C_2H_5 \quad + \quad NaBr$$

Because $RC\equiv C:^{(-)}\ Na^{(+)}$ is a very strong base (roughly a billion times stronger than NaOH), its use as a nucleophile in S_N2 reactions is limited to 1°-alkyl halides; 2° and 3°-alkyl halides undergo elimination by an E2 mechanism.

The enhanced acidity of terminal alkynes relative to alkanes also leads to metal exchange reactions when these compounds are treated with organolithium or Grignard reagents. This exchange, shown below in equation 1, can be interpreted as an acid-base reaction which, as expected, proceeds in the direction of the weaker acid and the weaker base. This factor clearly limits the usefulness of Grignard or lithium reagents when a terminal triple bond is present, as in equation 2.

1. $RC\equiv C\text{-}H + C_2H_5MgBr \text{ (in ether)}$

$$\longrightarrow RC\equiv C\text{-}MgBr \;+ C_2H_6$$

2. $HC\equiv C\text{-}CH_2CH_2Br + Mg \text{ (in ether)}$

$$\longrightarrow [HC\equiv C\text{-}CH_2CH_2MgBr]$$

$$\longrightarrow BrMgC\equiv C\text{-}CH_2CH_2H$$

The acidity of terminal alkynes also plays a role in product determination when vicinal (or geminal) dihalides undergo base induced bis-elimination reactions.

Chapter 8

Alkyl Halides or Haloalkanes

INTRODUCTION

Halogenoalkanes are compounds in which one or more hydrogen atoms in an alkane have been replaced by halogen atoms (fluorine, chlorine, bromine or iodine). Halogenoalkanes are also known as haloalkanes or alkyl halides.

For example:

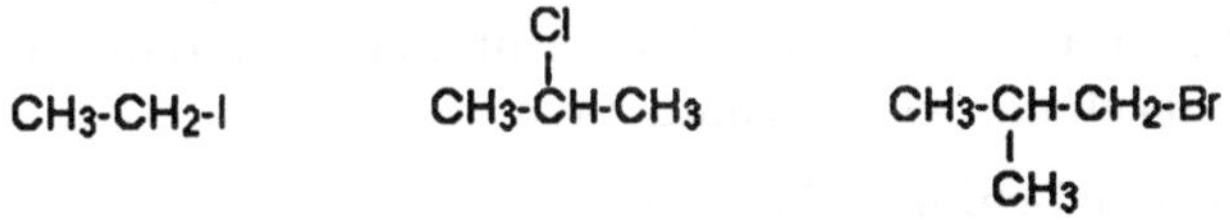

CH_3-CH_2-I — iodoethane

$CH_3-CH(Cl)-CH_3$ — 2-chloropropane

$CH_3-CH(CH_3)-CH_2-Br$ — 1-bromo-2methylpropane

NOMENCLATURE OF ALKYL HALIDES

We can use a common name approach for naming simple alkyl halides. We can attach the alkyl name of the hydrocarbon portion followed by the specific halide used using an ide ending. For example, CH_3-Cl would be called Methyl Chloride and CH_3-CH_2-Br would be called Ethyl Bromide. As the alkyl halide becomes more complex in its branching of the hydrocarbon or an alkyl halide with more than one halogen atom is used, then we have to resort to a more systematic approach and use the IUPAC rules.

The IUPAC rules for naming alkyl halides uses the same rules as is used for Alkanes if the halide is saturated, or for Alkenes or Alkynes if the halide is unsaturated. The halide(s) are treated as branched groups and are located on the continuous chain of carbons as you would locate and name any alkyl branch. For example, the structure in Fig 1-a below would be called 2,3-dichloro-5-methylheptane

```
                                      Br
                                      |
CH3-CH-CH-CH2                   CH3-CH-CH=CH
    |  |  |                               |
    Cl Cl CH-CH2-CH3                      CH—Cl
          |                               |
          CH3                             CH3
        (a)                          (b)
```

since the longest continuous chain of carbons with the halogen atoms attached is seven. We number from the end that will give us the lowest carbon number designators for the halide atoms and any alkyl groups.

The second structure (above) since the double bond is equal distant from either end and the halogen atoms are also equal distant from each end, we invoke a rule that says when this happens to take the branch that comes first alphabetically and make it attached to the lowest numbered carbon in the continuous chain. So we would have

2-bromo-5-chloro-3-hexene

DIFFERENT KINDS OF HALOGENOALKANES

Halogenoalkanes fall into different classes depending on how the halogen atom is positioned on the chain of carbon atoms. There are some chemical differences between the various types.

Primary Halogenoalkanes

In a primary (1°) halogenoalkane, the carbon which carries the halogen atom is only attached to one other alkyl group.

Some examples of primary halogenoalkanes include:

```
CH3—CH2—Br    CH3CH2—CH2—Cl    CH3CH—CH2—I
                                   |
                                   CH3
```

Notice that it doesn't matter how complicated the attached alkyl group is. In each case there is only *one linkage* to an alkyl group from the CH_2 group holding the halogen.

There is an exception to this. CH_3Br and the other methyl halides are often counted as primary halogenoalkanes even though there are *no* alkyl groups attached to the carbon with the halogen on it.

Secondary Halogenoalkanes

In a secondary (2°) halogenoalkane, the carbon with the halogen attached is joined directly to *two* other alkyl groups, which may be the same or different.

Examples:

$$\underset{\displaystyle \text{Br}}{CH_3-\underset{|}{CH}-CH_3} \qquad \underset{\displaystyle \text{Cl}}{CH_3-\underset{|}{CH}-CH_2CH_3}$$

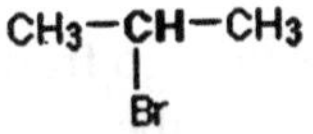

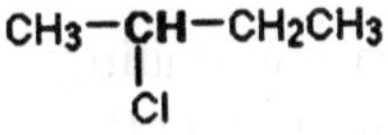

Tertiary Halogenoalkanes

In a tertiary (3°) halogenoalkane, the carbon atom holding the halogen is attached directly to *three* alkyl groups, which may be any combination of same or different.

Examples:

$$CH_3-\overset{\displaystyle CH_3}{\underset{\displaystyle Br}{\overset{|}{\underset{|}{C}}}}-CH_3 \qquad CH_3-\overset{\displaystyle CH_3}{\underset{\displaystyle Cl}{\overset{|}{\underset{|}{C}}}}-CH_2CH_3$$

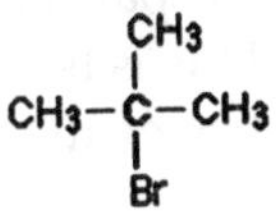

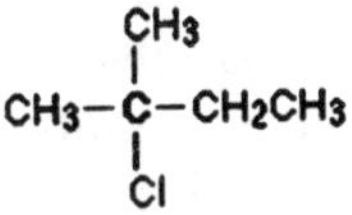

PHYSICAL PROPERTIES

Boiling Points

Methyl, ethyl and propyl halides have boiling points below room temperature (taken as being about 20°C). These are gases at room temperature. All the others you are likely to come across are liquids.

Remember:

- The only methyl halide which is a liquid is iodomethane;
- Chloroethane is a gas.

The patterns in boiling point reflect the patterns in intermolecular attractions.

Van der Waals Dispersion Forces

These attractions get stronger as the molecules get longer and have more electrons. That increases the sizes of the temporary dipoles that are set up.

This is why the boiling points increase as the number of carbon atoms in the chains increases. Look at the chart for a particular type of halide (a chloride, for example). Dispersion forces get stronger as you go from 1 to 2 to 3 carbons in the chain. It takes more energy to overcome them, and so the boiling points rise.

The increase in boiling point as you go from a chloride to a bromide to an iodide (for a given number of carbon atoms) is also because of the increase in number of electrons leading to larger dispersion forces. There are lots more electrons in, for example, iodomethane than there are in chloromethane.

Van der Waals Dipole-dipole Attractions

The carbon-halogen bonds (apart from the carbon-iodine bond) are polar, because the electron pair is pulled closer to the halogen atom than the carbon. This is because (apart from iodine) the halogens are more electronegative than carbon.

The electronegativity values are:

C	2.5	F	4.0
		Cl	3.0
		Br	2.8
		I	2.5

This means that in addition to the dispersion forces there will be forces due to the attractions between the permanent dipoles (except in the iodide case).

The size of those dipole-dipole attractions will fall as the bonds get less polar (as you go from chloride to bromide to iodide, for example). Nevertheless, the boiling points rise! This shows that the effect of the permanent dipole-dipole attractions is much less important than that of the temporary dipoles which cause the dispersion forces.

The large increase in number of electrons by the time you get to the iodide completely outweighs the loss of any permanent dipoles in the molecules.

Boiling Points of Some Isomers

	$CH_3-CH_2-CH_2-CH_2-Br$	$CH_3-CH_2-CH(Br)-CH_3$	$CH_3-C(CH_3)(Br)-CH_3$
B Pts:	375 K	364 K	346 K

The examples show that the boiling points fall as the isomers go from a primary to a secondary to a tertiary halogenoalkane. This is a simple result of the fall in the effectiveness of the dispersion forces.

The temporary dipoles are greatest for the longest molecule. The attractions are also stronger if the molecules can lie closely together. The tertiary halogenoalkane is very short and fat, and won't have much close contact with its neighbours.

Solubility

Solubility in Water

The halogenoalkanes are at best only very slightly soluble in water.

In order for a halogenoalkane to dissolve in water you have to break attractions between the halogenoalkane molecules (Van der Waals dispersion and dipole-dipole interactions) and break the hydrogen bonds between water molecules. Both of these cost energy.

Energy is released when new attractions are set up between the halogenoalkane and the water molecules. These will only be dispersion forces and dipole-dipole interactions. These aren't as strong as the original hydrogen bonds in the water, and so not as much energy is released as was used to separate the water molecules.

The energetics of the change are sufficiently "unprofitable" that very little dissolves.

Solubility in Organic Solvents

Halogenoalkanes tend to dissolve in organic solvents because the new intermolecular attractions have much the same strength as the ones being broken in the separate halogenoalkane and solvent.

Reactivity

The Importance of Bond Strengths

Bond strength falls as you go from C-F to C-I, and thus the carbon-fluorine bond is much stronger than the rest.

In order for anything to react with the halogenoalkanes, the carbon-halogen bond has got to be broken. Because that gets easier as you go from fluoride to chloride to bromide to iodide, the compounds get more reactive in that order.

Iodoalkanes are the most reactive and fluoroalkanes are the least.

The Influence of Bond Polarity

Of the four halogens, fluorine is the most electronegative and iodine the least. That means that the electron pair in the carbon-fluorine bond will be dragged most towards the halogen end.

The electronegativities of carbon and iodine are equal and so there is no separation of charge on the bond.

One of the important set of reactions of halogenoalkanes involves replacing the halogen by something else - substitution reactions. These reactions involve either:

- The carbon-halogen bond breaking to give positive and negative ions. The ion with the positively charged carbon atom then reacts with something either fully or slightly negatively charged.
- Something either fully or negatively charged attracted to the slightly positive carbon atom and pushing off the halogen atom.

You might have thought that either of these would be more effective in the case of the carbon-fluorine bond with the quite large amounts of positive and negative charge already present. But that's not so - quite the opposite is true!

The thing that governs the reactivity is the strength of the bonds which have to be broken. If is difficult to break a carbon-fluorine bond, but easy to break a carbon-iodine one.

Preparation

Halogenoalkanes can be made from the reaction between alkenes and hydrogen halides, but they are more commonly made by replacing the -OH group in an alcohol by a halogen atom.

Making Halogenoalkanes from Alcohols using Hydrogen Halides

The general reaction looks like this:

$$ROH + HX \longrightarrow RX + H_2O$$

Making Chloroalkanes

It is possible to make tertiary chloroalkanes successfully from the corresponding alcohol and concentrated hydrochloric acid, but to make primary or secondary ones you really need to use a different method - the reaction rates are too slow.

A tertiary chloroalkane can be made by shaking the corresponding alcohol with concentrated hydrochloric acid at room temperature.

$$CH_3\text{-}\underset{\displaystyle CH_3}{\overset{\displaystyle CH_3}{\underset{|}{\overset{|}{C}}}}\text{-}OH + HCl \longrightarrow CH_3\text{-}\underset{\displaystyle CH_3}{\overset{\displaystyle CH_3}{\underset{|}{\overset{|}{C}}}}\text{-}Cl + H_2O$$

Making Bromoalkanes

Rather than using hydrobromic acid, you usually treat the alcohol with a mixture of sodium or potassium bromide and concentrated sulphuric acid. This produces hydrogen bromide which reacts with the alcohol. The mixture is warmed to distil off the bromoalkane.

$$CH_3\text{-}CH_2\text{-}OH + HBr \longrightarrow CH_3\text{-}CH_2\text{-}Br + H_2O$$

Making Iodoalkanes

In this case the alcohol is reacted with a mixture of sodium or potassium iodide and concentrated phosphoric(V) acid, H_3PO_4, and the iodoalkane is distilled off. The mixture of the iodide and phosphoric(V) acid produces hydrogen iodide which reacts with the alcohol.

$$CH_3\text{-}CH_2\text{-}OH + HI \longrightarrow CH_3\text{-}CH_2\text{-}I + H_2O$$

Phosphoric(V) acid is used instead of concentrated sulphuric acid because sulphuric acid oxidizes iodide ions to iodine and produces hardly any hydrogen iodide. A similar thing happens to some extent with bromide ions in the preparation of bromoalkanes, but not enough to get in the way of the main reaction.

Making Halogenoalkanes from Alcohols using Phosphorus Halides

Making Chloroalkanes

Chloroalkanes can be made by reacting an alcohol with liquid phosphorus(III) chloride, PCl_3.

They can also be made by adding solid phosphorus(V) chloride, PCl_5 to an alcohol.

This reaction is violent at room temperature, producing clouds of hydrogen chloride gas. It isn't a good choice as a way of making halogenoalkanes, although it is used as a test for -OH groups in organic chemistry.

$$CH_3\text{-}CH_2\text{-}CH_2\text{-}OH + PCl_5 \longrightarrow CH_3\text{-}CH_2\text{-}CH_2\text{-}Cl + POCl_3 + HCl$$

There are also side reactions involving the $POCl_3$ reacting with the alcohol.

Making Bromoalkanes and Iodoalkanes

These are both made in the same general way. Instead of using phosphorus(III) bromide or iodide, the alcohol is heated under reflux with a mixture of red phosphorus and either bromine or iodine.

The phosphorus first reacts with the bromine or iodine to give the phosphorus(III) halide.

$$2P + 3Br_2 \longrightarrow 2PBr_3$$

$$2P + 3I_2 \longrightarrow 2PI_3$$

These then react with the alcohol to give the corresponding halogenoalkane which can be distilled off.

$$3CH_3\text{-}CH_2\text{-}OH + PBr_3 \longrightarrow 3CH_3\text{-}CH_2\text{-}Br + H_3PO_3$$

$$3CH_3\text{-}CH_2\text{-}OH + PI_3 \longrightarrow 3CH_3\text{-}CH_2\text{-}I + H_3PO_3$$

MAKING BROMOETHANE IN THE LAB

Making Impure Bromoethane

Concentrated sulphuric acid is added slowly with lots of shaking and cooling to some ethanol in a flask, and then solid potassium bromide is added. The flask is then connected to a condenser so that the bromoethane formed can be distilled off.

$$H_2SO_4 + KBr \longrightarrow KHSO_4 + HBr$$

$$CH_3\text{-}CH_2\text{-}OH + HBr \longrightarrow CH_3\text{-}CH_2\text{-}Br + H_2O$$

Bromoethane has a low boiling point but is denser than water and almost insoluble in it. To prevent it from evaporating, it is often collected under water in a flask surrounded by ice. Sometimes it is simply collected in a tube surrounded by ice without any water.

The reaction flask is heated gently until no more droplets of bromoethane collect.

Purifying the Bromoethane

Impurities in the bromoethane include:

- Hydrogen bromide (although most of that will dissolve in the water if you are collecting the bromoethane under water);

- Bromine - from the oxidation of bromide ions by the concentrated sulphuric acid;
- Sulphur dioxide - formed when concentrated sulphuric acid oxidizes the bromide ions;
- Unreacted ethanol;
- Ethoxyethane (diethyl ether) - formed by a side reaction between the ethanol and the concentrated sulphuric acid.

The Purification Sequence

Stage 1

If you have collected the bromoethane under water, transfer the contents of the collection flask to a separating funnel. Otherwise, pour the impure bromoethane into the separating funnel, add some water and shake it.

Pour off and keep the bromoethane layer.

The water you discard will contain almost all of the hydrogen bromide, and quite a lot of any bromine, sulphur dioxide and ethanol present as impurities.

Stage 2

To get rid of any remaining acidic impurities (including the bromine and sulphur dioxide), return the bromoethane to the separating funnel and shake it with either sodium carbonate or sodium hydrogencarbonate solution.

This reacts with any acids present liberating carbon dioxide and forming soluble salts.

Separate and retain the lower bromoethane layer as before.

Stage 3

Now wash the bromoethane with water in a separating funnel to remove any remaining inorganic impurities (excess sodium carbonate solution, etc). This time, transfer the lower bromoethane layer to a dry test tube.

Stage 4

Add some anhydrous calcium chloride to the tube, shake well and leave to stand. The anhydrous calcium chloride is a drying agent and removes any remaining water. It also absorbs ethanol, and so any remaining ethanol may be removed as well (depending on how much calcium chloride you use).

Stage 5

Transfer the dry bromoethane to a distillation flask and fractionally distil it, collecting what distils over at between 35 and 40°C.

In principle, this should remove any remaining organic impurities. In practice, though, any ethoxyethane (which is perhaps the most likely impurity left at this stage) has a boiling point very, very close to that of bromoethane. It is unlikely that you will be able to separate the two.

If there is any ethanol left which hadn't been absorbed by the calcium chloride, that would certainly be removed because its boiling point is much higher.

REACTIONS

WITH HYDROXIDE IONS

Substitution or Elimination?

There are two different sorts of reaction that you can get depending on the conditions used and the type of halogenoalkane. Primary, secondary and tertiary halogenoalkanes behave differently in this respect.

Substitution Reactions

In a substitution reaction, the halogen atom is replaced by an -OH group to give an alcohol.

For example:

$$CH_3\underset{\underset{Br}{|}}{C}HCH_3 + NaOH \longrightarrow CH_3\underset{\underset{OH}{|}}{C}HCH_3 + NaBr$$

Or, as an ionic equation:

$$CH_3\underset{\underset{Br}{|}}{C}HCH_3 + OH^- \longrightarrow CH_3\underset{\underset{OH}{|}}{C}HCH_3 + Br^-$$

In the example, 2-bromopropane is converted into propan-2-ol.

The halogenoalkane is heated under reflux with a solution of sodium or potassium hydroxide. Heating under reflux means heating with a condenser placed vertically in the flask to prevent loss of volatile substances from the mixture.

The solvent is usually a 50/50 mixture of ethanol and water, because everything will dissolve in that. The halogenoalkane is insoluble in water. If you used water alone as the solvent, the halogenoalkane and the sodium hydroxide solution wouldn't mix and the reaction could only happen where the two layers met.

Elimination Reactions

Halogenoalkanes also undergo elimination reactions in the presence of sodium or potassium hydroxide.

$$CH_3\underset{\underset{Br}{|}}{C}HCH_3 + NaOH \longrightarrow CH_2{=}CHCH_3 + NaBr + H_2O$$

The 2-bromopropane has reacted to give an alkene - propene.

Notice that a hydrogen atom has been removed from one of the end carbon atoms together with the bromine from the centre one. In all simple elimination reactions the things being removed are on adjacent carbon atoms, and a double bond is set up between those carbons.

The halogenoalkane is heated under reflux with a concentrated solution of sodium or potassium hydroxide in ethanol. Propene is formed and, because this is a gas, it passes through the condenser and can be collected.

What Decides whether you Get Substitution or Elimination?

The reagents you are using are the same for both substitution or elimination — the halogenoalkane and either sodium or potassium hydroxide solution. In all cases, you will get a mixture of both reactions happening - some substitution and some elimination. What you get most of depends on a number of factors.

The Type of Halogenoalkane

This is the most important factor.

Type of Halogenoalkane	*Substitution or Elimination?*
primary	mainly substitution
secondary	both substitution and elimination
tertiary	mainly elimination

For example, whatever you do with tertiary halogenoalkanes, you will tend to get mainly the elimination reaction, whereas with primary ones you will tend to get mainly substitution. However, you can influence things to some extent by changing the conditions.

The Solvent

The proportion of water to ethanol in the solvent matters.

- Water encourages substitution.
- Ethanol encourages elimination.

The Temperature

Higher temperatures encourage elimination.

Concentration of the Sodium or Potassium Hydroxide Solution

Higher concentrations favour elimination.

In Summary

For a given halogenoalkane, to favour *elimination* rather than substitution, use:

- Higher temperatures
- A concentrated solution of sodium or potassium hydroxide
- Pure ethanol as the solvent

To favour substitution rather than elimination, use:

- Lower temperatures
- More dilute solutions of sodium or potassium hydroxide
- More water in the solvent mixture

WITH CYANIDE IONS

Replacing a Halogen by -CN

If a halogenoalkane is heated under reflux with a solution of sodium or potassium cyanide in ethanol, the halogen is replaced by a -CN group and a nitrile is produced. Heating under reflux means heating with a condenser placed vertically in the flask to prevent loss of volatile substances from the mixture.

The solvent is important. If water is present you tend to get substitution by -OH instead of -CN.

For example, using 1-bromopropane as a typical primary halogenoalkane:

$$CH_3CH_2CH_2Br + CN^- \longrightarrow CH_3CH_2CH_2CN + Br$$

You could write the full equation rather than the ionic one, but it slightly obscures what's going on:

$$CH_3CH_2CH_2Br + KCN \longrightarrow CH_3CH_2CH_2CN + KBr$$

The bromine (or other halogen) in the halogenoalkane is simply replaced by a -CN group - hence a substitution reaction. In this example, butanenitrile is formed.

Secondary and tertiary halogenoalkanes behave similarly, although the mechanism will vary depending on which sort of halogenoalkane you are using.

This reaction with cyanide ions is a useful way of lengthening carbon chains. For example, in the equations above, you start with a 3-carbon chain and end up with a 4-carbon chain. There aren't very many simple ways of making new carbon-carbon bonds.

It is fairly easy to change the -CN group at the end of the new chain into other groups.

WITH AMMONIA

Reaction Details and Products

The halogenoalkane is heated with a concentrated solution of ammonia in ethanol. The reaction is carried out in a sealed tube. You couldn't heat this mixture under reflux, because the ammonia would simply escape up the condenser as a gas.

We'll talk about the reaction using 1-bromoethane as a typical primary halogenoalkane. There is no difference in the details of the reaction if you chose a secondary or tertiary halogenoalkane instead. The equations would just look more complicated than they already are!

You get a series of amines formed together with their salts. The reactions happen one after another.

Making a Primary Amine

The reaction happens in two stages. In the first stage, a salt is formed - in this case, ethylammonium bromide. This is just like ammonium bromide, except that one of the hydrogens in the ammonium ion is replaced by an ethyl group.

$$CH_3CH_2Br + NH_3 \longrightarrow CH_3CH_2CH_3CN + Br$$

There is then the possibility of a reversible reaction between this salt and excess ammonia in the mixture.

$$CH_3CH_2NH_3 + Br + NH_3 \rightleftharpoons CH_3CH_2NH_2 + NH_4 + Br$$

The ammonia removes a hydrogen ion from the ethylammonium ion to leave a primary amine - ethylamine.

The more ammonia there is in the mixture, the more the forward reaction is favoured.

Making a Secondary Amine

The reaction doesn't stop at a primary amine. The ethylamine also reacts with bromoethane - in the same two stages as before.

In the first stage, you get a salt formed - this time, diethylammonium bromide. Think of this as ammonium bromide with two hydrogens replaced by ethyl groups.

$$CH_3CH_2Br + CH_3CH_2NH_2 \longrightarrow (CH_3CH_2)_2NH_2^+ \; Br^-$$

There is again the possibility of a reversible reaction between this salt and excess ammonia in the mixture.

$$(CH_3CH_2)_2NH_2^+ \; Br^- + NH_3 \rightleftharpoons (CH_3CH_2)_2NH + NH_4^+ Br^-$$

The ammonia removes a hydrogen ion from the diethylammonium ion to leave a secondary amine - diethylamine. A secondary amine is one which has two alkyl groups attached to the nitrogen.

Making a Tertiary Amine

The diethylamine also reacts with bromoethane - in the same two stages as before.

In the first stage, you get triethylammonium bromide.

$$CH_3CH_2Br + (CH_3CH_2)_2NH \longrightarrow (CH_3CH_2)_3NH^+ \; Br^-$$

There is again the possibility of a reversible reaction between this salt and excess ammonia in the mixture.

$$(CH_3CH_2)_3NH^+ \; Br^- + NH_3 \rightleftharpoons (CH_3CH_2)_3N + NH_4^+ Br^-$$

The ammonia removes a hydrogen ion from the triethylammonium ion to leave a tertiary amine - triethylamine. A tertiary amine is one which has three alkyl groups attached to the nitrogen.

Making a Quaternary Ammonium Salt

The final stage! The triethylamine reacts with bromoethane to give tetraethylammonium bromide - a quaternary ammonium salt (one in which all four hydrogens have been replaced by alkyl groups).

$$CH_3CH_2Br + CH_3CH_2-\overset{\displaystyle CH_3CH_2}{\underset{\displaystyle CH_3CH_2}{N}} \longrightarrow CH_3CH_2-\overset{\displaystyle CH_3CH_2}{\underset{\displaystyle CH_3CH_2}{N^+}}-CH_2CH_3 \quad Br^-$$

This time there isn't any hydrogen left on the nitrogen to be removed. The reaction stops here.

What do you Actually Get if you React Bromoethane with Ammonia?

Whatever you do, you get a mixture of all of the products (including both amines and their salts) shown above.

To get mainly the quaternary ammonium salt, you can use a large excess of bromoethane. If you look at the reactions going on, each one needs additional bromoethane. If you provide enough, then the chances are that the reaction will go to completion, given enough time.

On the other hand, if you use a very large excess of ammonia, the chances are always greatest that a bromoethane molecule will hit an ammonia molecule rather than one of the amines being formed. That will help to prevent the formation of secondary (etc) amines.

WITH SILVER NITRATE SOLUTION

Testing for Halogenoalkanes

Silver nitrate solution can be used to find out which halogen is present in a suspected halogenoalkane. The most effective way is to do a substitution reaction which turns the halogen into a halide ion, and then to test for that ion with silver nitrate solution.

The halogenoalkane is warmed with some sodium hydroxide solution in a mixture of ethanol and water.

Everything will dissolve in this mixture and so you can get a good reaction.

The halogen atom is displaced as a halide ion:

There is no need to make this reaction go to completion. The silver nitrate test is sensitive enough to detect fairly small concentrations of halide ions.

The mixture is acidified by adding dilute nitric acid. This prevents unreacted hydroxide ions reacting with the silver ions. Then silver nitrate solution is added.

Various precipitates may be formed from the reaction between the silver and halide ions:

$$Ag^+(aq) + Hal^-(aq) \longrightarrow AgHal(s)$$

Ion Present	*Observation*
Cl^-	white precipitate
Br^-	very pale cream precipitate
I^-	very pale yellow precipitate

Confirming the Precipitates

It is actually quite difficult to distinguish between these colours, especially if there isn't much precipitate. You can sort out which precipitate you have by adding *ammonia solution.*

Original Precipitate	*Observation*
AgCl	precipitate dissolves to give a colourless solution
AgBr	precipitate is almost unchanged using dilute ammonia solution, but dissolves in concentrated ammonia solution to give a colourless solution
AgI	precipitate is insoluble in ammonia solution of any concentration

Comparing Halogenoalkane Reactivities

In this case, various halogenoalkanes are treated with a solution of silver nitrate in a mixture of ethanol and water. Nothing else is added. After varying lengths of time precipitates appear as halide ions (produced from reactions of the halogenoalkanes) react with the silver ions present.

As long as you are doing everything under controlled conditions (same amounts of everything, same temperature and so on), the time taken gives a good guide to the reactivity of the halogenoalkanes - the quicker the precipitate appears, the more reactive the halogenoalkane.

The halide ion is formed in one of two ways, depending on the type of halogenoalkane you have present - primary, secondary or tertiary.

For a primary halogenoalkane, the main reaction is one between the halogenoalkane and water in the solvent.

$$\text{R-Hal} + H_2O \longrightarrow \text{R-OH} + H^+ + Hal^-$$

A tertiary halogenoalkane ionises to a very small extent of its own accord.

$$\text{R-Hal} \rightleftharpoons R^+ + Hal^-$$

Secondary halogenoalkanes do a bit of both of these.

Comparing the Reaction rates as you Change the Halogen

You would have to keep the type of halogenoalkane (primary, secondary or tertiary) constant, but vary the halogen. You might, for example, compare the times taken to produce a precipitate from this series of primary halogenoalkanes:

$$\text{R-Hal} + OH^- \longrightarrow \text{R-OH} + Hal^-$$

CH_3-CH_2-CH_2-CH_2-Cl — 1-Chlorobutane

CH_3-CH_2-CH_2-CH_2-Br — 1-bromobutane

CH_3-CH_2-CH_2-CH_2- — 1-iodobutane

Obviously, the time taken for a precipitate of silver bromide to appear will depend on how much of everything you use and the temperature at which the reaction is carried out. But the pattern of results is always the same.

For example:

- A primary iodo compound produces a precipitate quite quickly.

- A primary bromo compound takes longer to give a precipitate.
- A primary chloro compound probably won't give any precipitate.
- The order of reactivity reflects the strengths of the carbon-halogen bonds. The carbon-iodine bond is the weakest and the carbon-chlorine the strongest of the three bonds. In order for a halide ion to be produced, the carbon-halogen bond has to be broken. The weaker the bond, the easier that is.

Comparing the Reaction Rates of Primary, Secondary and Tertiary Halogenoalkanes

You would need to keep the halogen atom constant. It is common to use bromides because they have moderate reaction rates. You could, for example, compare the reactivity of these compounds:

$CH_3\text{-}CH_2\text{-}CH_2\text{-}CH_2\text{-}Br$	$CH_3\text{-}CH_2\text{-}CH(Br)\text{-}CH_3$	$CH_3\text{-}C(CH_3)(Br)\text{-}CH_3$
1-bromobutane (primary)	2-bromobutane (secondary)	2-bromo-2-methylpropane (tertiary)

Again, the actual times taken will vary with reaction conditions, but the pattern will always be the same.

For example:

- The tertiary halide produces a precipitate almost instantly.
- The secondary halide gives a slight precipitate after a few seconds. The precipitate thickens up with time.
- The primary halide may take considerably longer to produce a precipitate.

It is more difficult to explain the reason for this, because it needs a fairly intimate knowledge of the mechanisms involved in the reactions. It reflects the change in the way that the halide ion is produced as you go from primary to secondary to tertiary halogenoalkanes.

AN INTRODUCTION TO GRIGNARD REAGENTS

A Grignard reagent has a formula RMgX where X is a halogen, and R is an alkyl or aryl (based on a benzene ring) group. For the purposes of this section we shall take R to be an alkyl group.

A typical Grignard reagent might be CH_3CH_2MgBr.

The Preparation of a Grignard Reagent

Grignard reagents are made by adding the halogenoalkane to small bits of magnesium in a flask containing ethoxyethane (commonly called diethyl ether or just "ether"). The flask is fitted with a reflux condenser, and the mixture is warmed over a water bath for 20 - 30 minutes.

$$CH_3CH_2Br + Mg \xrightarrow{\text{ethoxyethane}} CH_3CH_2MgBr$$

Everything must be perfectly dry because Grignard reagents react with water.

Any reactions using the Grignard reagent are carried out with the mixture produced from this reaction. You can't separate it out in any way.

REACTIONS OF GRIGNARD REAGENTS

Grignard Reagents and Water

Grignard reagents react with water to produce alkanes. This is the reason that everything has to be very dry during the preparation above.

For example:

$$CH_3CH_2MgBr + H_2O \longrightarrow CH_3CH_3 + Mg(OH)Br$$

The inorganic product, Mg(OH)Br, is referred to as a "basic bromide". You can think of it as a sort of half-way stage between magnesium bromide and magnesium hydroxide.

$$CH_3CH_2MgBr + H_2O \longrightarrow CH_3CH_3 + Mg(OH)Br$$

GRIGNARD REAGENTS AND CARBON DIOXIDE

Grignard reagents react with carbon dioxide in two stages. In the first, you get an addition of the Grignard reagent to the carbon dioxide.

Dry carbon dioxide is bubbled through a solution of the Grignard reagent in ethoxyethane, made as described above.

For example:

$$CH_3CH_2MgBr + O{=}C{=}O \longrightarrow CH_3CH_2C(=O)\text{O-MgBr}$$

The CH_3CH_2MgBr adds an across this double bond

The product is then hydrolyzed (reacted with water) in the presence of a dilute acid. Typically, you would add dilute sulphuric acid or dilute hydrochloric acid to the solution formed by the reaction with the CO_2.

A carboxylic acid is produced with one more carbon than the original Grignard reagent.

The usually quoted equation is (without the red bits):

dilute sulphuric acid

$$CH_3CH_2C(=O)\text{O-MgBr} + H_2O \xrightarrow{H_3O^+} CH_3CH_2C(=O)\text{O-H} + Mg(OH)Br$$

This reacts with the dilute sulphuric acid to give ordinary Mg^{2+} (aq) ions, bromide ions and water

Almost all sources quote the formation of a basic halide such as Mg(OH)Br as the other product of the reaction. That's actually misleading because these compounds react with dilute acids. What you end up with would be a mixture of ordinary hydrated magnesium ions, halide ions and sulphate or chloride ions - depending on which dilute acid you added.

Grignard Reagents and Carbonyl Compounds

The reactions between the various sorts of carbonyl compounds and Grignard reagents can look quite complicated, but in fact they all react in the same way - all that changes are the groups attached to the carbon-oxygen double bond.

It is much easier to understand what is going on by looking closely at the general case and then slotting in the various real groups as and when you need to.

The reactions are essentially identical to the reaction with carbon dioxide - all that differs is the nature of the organic product.

In the first stage, the Grignard reagent adds across the carbon-oxygen double bond:

$$CH_3CH_2MgBr + R{-}C({=}O){-}R' \longrightarrow CH_3CH_2{-}C(R)(R'){-}O{-}MgBr$$

Dilute acid is then added to this to hydrolyze it. (I am using the normally accepted equation ignoring the fact that the Mg(OH)Br will react further with the acid.)

$$CH_3CH_2{-}C(R)(R'){-}O{-}MgBr + H_2O \xrightarrow{H_3O^+} CH_3CH_2{-}C(R)(R'){-}OH + Mg(OH)Br$$

An alcohol is formed. One of the key uses of Grignard reagents is the ability to make complicated alcohols easily.

What sort of alcohol you get depends on the carbonyl compound you started with.

The Reaction between Grignard Reagents and Methanal

In methanal, both R groups are hydrogen. Methanal is the simplest possible aldehyde.

$$H{-}C({=}O){-}H$$

methanal

Assuming that you are starting with CH_3CH_2MgBr and using the general equation above, the alcohol you get always has the form:

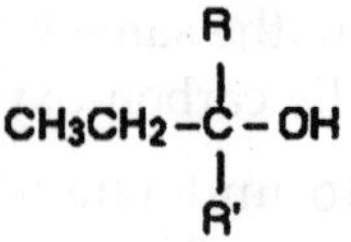

$$CH_3CH_2-\underset{R'}{\overset{R}{\underset{|}{\overset{|}{C}}}}-OH$$

Since both R groups are hydrogen atoms, the final product will be:

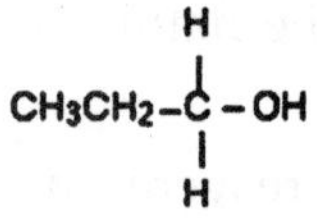

$$CH_3CH_2-\underset{H}{\overset{H}{\underset{|}{\overset{|}{C}}}}-OH \quad \text{or} \quad CH_3CH_2CH_2OH$$

a primary alcohol

A primary alcohol is formed. A primary alcohol has only one alkyl group attached to the carbon atom with the -OH group on it.

You could obviously get a different primary alcohol if you started from a different Grignard reagent.

The Reaction between Grignard Reagents and other Aldehydes

The next biggest aldehyde is ethanal. One of the R groups is hydrogen and the other CH_3.

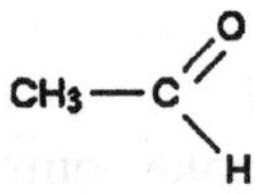

$$CH_3-C(=O)H$$

ethanal

Again, think about how that relates to the general case. The alcohol formed is:

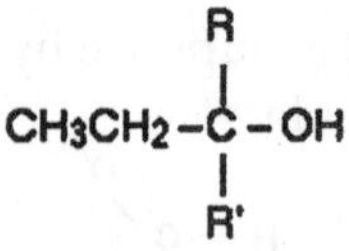

$$CH_3CH_2-\underset{R'}{\overset{R}{\underset{|}{\overset{|}{C}}}}-OH$$

So this time the final product has one CH_3 group and one hydrogen attached:

$$CH_3CH_2-\overset{\displaystyle CH_3}{\underset{\displaystyle H}{\overset{|}{\underset{|}{C}}}}-OH \quad \text{or} \quad CH_3CH_2\overset{\displaystyle CH_3}{\overset{|}{C}}HOH$$

a secondary alcohol

A secondary alcohol has two alkyl groups (the same or different) attached to the carbon with the -OH group on it.

You could change the nature of the final secondary alcohol by either:

- Changing the nature of the Grignard reagent - which would change the CH_3CH_2 group into some other alkyl group;
- Changing the nature of the aldehyde - which would change the CH_3 group into some other alkyl group.

The Reaction between Grignard Reagents and Ketones

Ketones have two alkyl groups attached to the carbon-oxygen double bond. The simplest one is propanone.

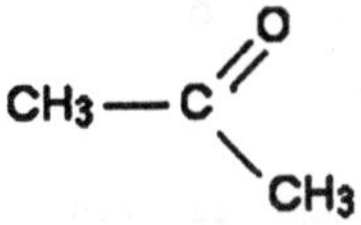

$$CH_3-C(=O)-CH_3$$

propanone

This time when you replace the R groups in the general formula for the alcohol produced you get a tertiary alcohol.

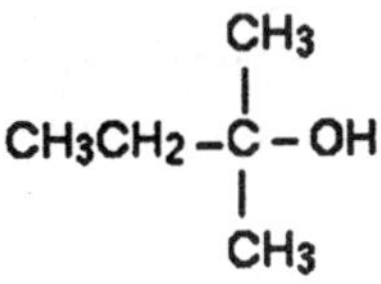

$$CH_3CH_2-\overset{\displaystyle CH_3}{\underset{\displaystyle CH_3}{\overset{|}{\underset{|}{C}}}}-OH$$

a tertiary alcohol

A tertiary alcohol has three alkyl groups attached to the carbon with the -OH attached. The alkyl groups can be any combination of same or different.

You could ring the changes on the product by

- Changing the nature of the Grignard reagent - which would change the CH_3CH_2 group into some other alkyl group;
- Changing the nature of the ketone - which would change the CH_3 groups into whatever other alkyl groups you choose to have in the original ketone.

Why do Grignard Reagents React with Carbonyl Compounds?

The bond between the carbon atom and the magnesium is polar. Carbon is more electronegative than magnesium, and so the bonding pair of electrons is pulled towards the carbon.

That leaves the carbon atom with a slight negative charge.

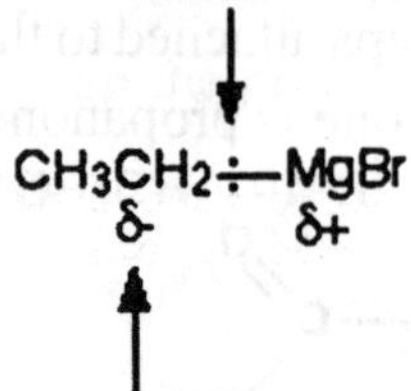

The carbon-oxygen double bond is also highly polar with a significant amount of positive charge on the carbon atom. The Grignard reagent can therefore serve as a *nucleophile* because of the attraction between the slight negativeness of the carbon atom in the Grignard reagent and the positiveness of the carbon in the carbonyl compound.

Chapter 9

Alcohols

INTRODUCTION

Alcohols are a family of organic compounds that share a common chemically active group called the hydroxyl group(-O-H). These compounds are formed by replacing one or more hydrogen atoms in an alkane by an -OH group. The hydroxyl group is not to be confused with the hydroxide ion which is a negatively charged Oxygen bonded to a hydrogen. The hydroxyl group is electrically neutral. It represents the position in an alcohol molecule where the chemical change takes place.

Monohydric alcohols are alcohols with only one hydroxyl group attached to one of its carbons. Polyhydric alcohols would be alcohols that have two or more hydroxyl groups attached to the molecule. There are "diols", "triols", etc. The general symbol to represent any monohydric alcohol is R-O-H.

For example:

CH_3-CH_2-OH	CH_3-CH(OH)-CH_3	CH_3-CH(CH_3)-CH_2-OH
ethanol	propan-2-ol	2-methylpropan-1-ol

NOMENCLATURE OF ALCOHOLS

Simple alcohols can be named using the common name approach of identifying and naming the alkyl portion and follow it with the word alcohol. For example, CH_3-OH could be called methyl alcohol since the CH_3 group is called a Methyl group. $(CH_3)_2CH$-OH could be called Isopropyl Alcohol since the hydrocarbon portion is the isopropyl group. This method works fine as long as you have simple alcohols. When the alcohols becomes more complex and more branched, then this method becomes much more difficult. The IUPAC have come up with a set of rules that are used to name any alcohol regardless of its complexity. These rules are summarized as follows:

1. Identify the longest continuous chain of carbons in which the Hydroxyl group(s) are attached to one of the carbons in the chain.
2. Number the chain so that the hydroxyl group(s) are attached to the lowest numbered carbons.
3. Identify and locate the other branches on the chain so that they are named alphabetically and their carbon number is hyphenated onto the front of the name. If more than one of the same group is present use the greek prefix attached to the branch name. (di=2, tri=3, etc).
4. After all the branches have been named and located then attach the carbon number that is attached to the hydroxyl group onto the alkane name associated with the number of carbons found in the continuous chain in step 1.
5. drop the "e" on the alkane name and attach the characteristic IUPAC ending ol to the rest of the alkane name.
6. For polyhydric alcohols you would have to locate two or more carbons which the Hydroxyl groups were attached and hyphenate those carbon numbers to the front of the alkane name and attach "diol" if two are involved, "triol" if three hydroxyl groups were involved, etc.

CH_3-CH(CH_3)-CH_2-CH_2-CH(Cl)-CH_2-OH

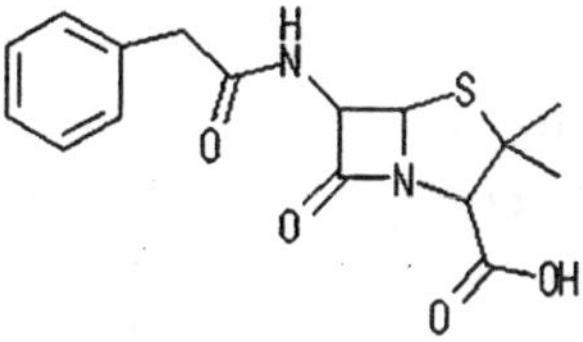

6-pheny lacety lamino-3-carboxyl-2,2-dimethyl-1
-thia-4-azabicyclo[3.2.0]heptan-5-one

Fig. 1a

In Fig 1a above we will use the IUPAC steps to name it.

1. Identify the longest continuous chain with the Hydroxyl group involved. That would be six carbons in length
2. Number the carbons in the chain so that the Hydroxyl group is attached to the lowest numbered carbon in the chain. That would be to number from right to left
3. Identify, locate, and name alphabetically the other branches. There is a Chloro group on the number 2 carbon and a methyl group on the number 5 carbon so the first part of the name would be:

2-Chloro-5 methyl

4. Determine the alkane name associated with six carbons which would be hexane.
5. Locate the position of the hydroxyl group which is on carbon number 1.
6. Hyphenate the carbon number onto the alkane name drop the "e" and add the characteristic "ol" ending.

2-Chloro-5-methyl-1-hexanol

CH_3-CH(OH)-CH(C_2H_5)-CH_2-CH(OH)-CH_2-CH_2-CH_2-CH_2-Br

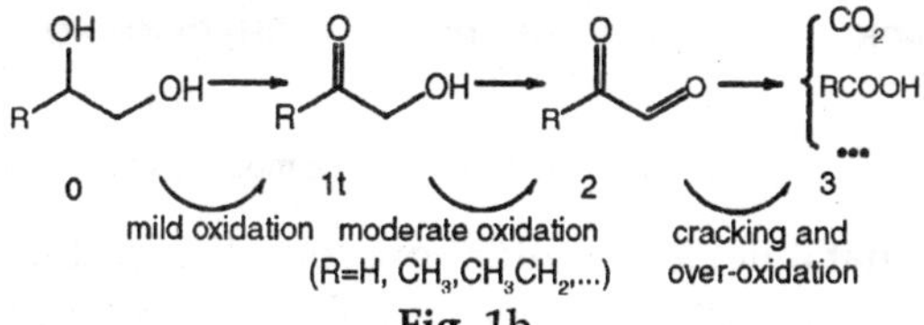

Fig. 1b

In Fig 1-b above, we have another alcohol that is a polyhydric alcohol. Let's use the IUPAC rules to name it.

1. Determine the longest continuous carbon chain in which both Hydroxyl groups are involved. That would be nine carbons in length.
2. Number the carbons in the chain so that the Hydroxyl groups are attached to the lowest numbered carbons. That would be to number from left to right.
3. Identify the other branches and prefix the carbon numbers to the branch groups. There is a Bromo group on carbon # 9 and an ethyl group on carbon # 3. So naming alphabetically we would have:

9-Bromo-3-ethyl

4. Attach the alkane name for nine carbons which would be Nonane
5. Hyphenate the carbon numbers that have the Hydroxyl groups attached and attach the diol ending to the alkane name

9-Bromo-3-ethyl-2,5-nonanediol

DIFFERENT KINDS OF ALCOHOLS

Alcohols fall into different classes depending on how the -OH group is positioned on the chain of carbon atoms. There are some chemical differences between the various types.

Primary Alcohols

In a primary (1°) alcohol, the carbon which carries the -OH group is only attached to one alkyl group.

Some examples of primary alcohols include:

CH_3-CH_2-OH	CH_3-CH_2-CH_2-OH	CH_3-CH(-CH_3)-CH_2-OH
ethanol	propan-1-ol	2-methylpropan-1-ol

Notice that it doesn't matter how complicated the attached alkyl group is. In each case there is only *one linkage* to an alkyl group from the CH_2 group holding the -OH group.

There is an exception to this. Methanol, CH_3OH, is counted as a primary alcohol even though there are *no* alkyl groups attached to the carbon with the -OH group on it.

Secondary Alcohols

In a secondary (2°) alcohol, the carbon with the -OH group attached is joined directly to *two* alkyl groups, which may be the same or different.

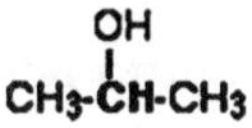

$$CH_3-\underset{}{\overset{OH}{\overset{|}{C}}}H-CH_3 \qquad CH_3-\overset{OH}{\overset{|}{C}}H-CH_2-CH_3 \qquad CH_3-CH_2-\overset{OH}{\overset{|}{C}}H-CH_2-CH_3$$

propan-2-ol 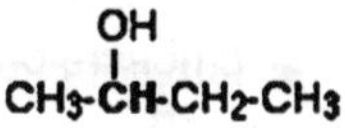 butan-2-ol 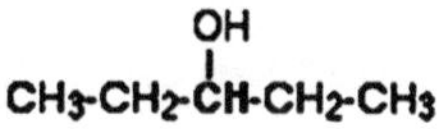 pentan-3-ol

Tertiary Alcohols

In a tertiary (3°) alcohol, the carbon atom holding the -OH group is attached directly to *three* alkyl groups, which may be any combination of same or different.

Examples:

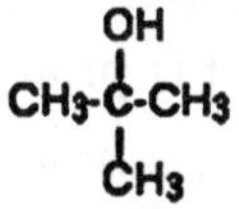

$$CH_3-\underset{\underset{CH_3}{|}}{\overset{OH}{\overset{|}{C}}}-CH_3 \qquad CH_3-CH_2-\underset{\underset{CH_3}{|}}{\overset{OH}{\overset{|}{C}}}-CH_3$$

2-methylpropan-2-ol 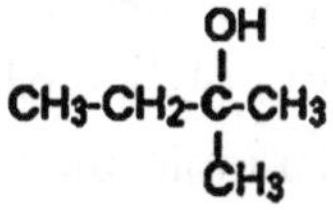 2-methylbutan-2-ol

PREPARATION

From Alkenes

The preparation of ethanol from ethene

Ethanol is manufactured by reacting ethene with steam. The catalyst used is solid silicon dioxide coated with phosphoric(V) acid. The reaction is reversible.

$$CH_2=CH_{2(g)} + H_2O_{(g)} \xrightleftharpoons{H_3PO_4} CH_3CH_2OH_{(g)}$$

Only 5% of the ethene is converted into ethanol at each pass through the reactor. By removing the ethanol from the equilibrium mixture and recycling the ethene, it is possible to achieve an overall 95% conversion.

Preparation of other Alcohols from Alkenes

Some - but not all - other alcohols can be made by similar reactions. The catalyst used and the reaction conditions will vary from alcohol to alcohol. The reason that there is a problem with some alcohols is well illustrated with trying to make an alcohol from propene, $CH_3CH=CH_2$.

In principle, there are two different alcohols which might be formed:

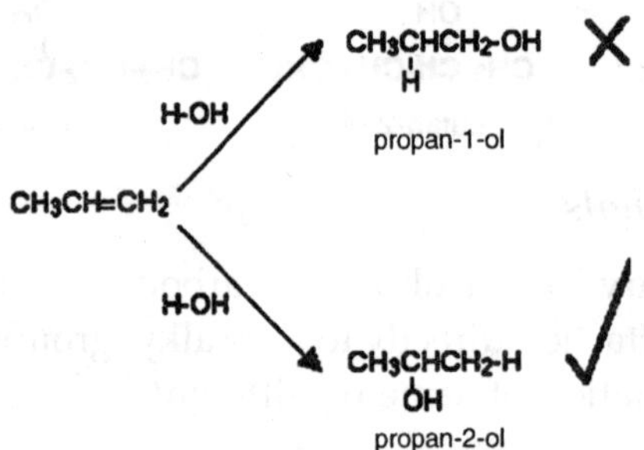

You might expect to get either propan-1-ol or propan-2-ol depending on which way around the water adds to the double bond. In practice what you get is propan-2-ol.

If you add a molecule H-X across a carbon-carbon double bond, the hydrogen nearly always gets attached to the carbon with the most hydrogens on it already - in this case the CH_2 rather than the CH.

The effect of this is that there are bound to be some alcohols which it is impossible to make by reacting alkenes with steam because the addition would be the wrong way around.

MAKING ETHANOL BY FERMENTATION

This method *only* applies to ethanol. You can't make any other alcohol this way.

Process of Making

The starting material for the process varies widely, but will normally be some form of starchy plant material such as maize (US: corn), wheat, barley or potatoes.

Starch is a complex carbohydrate, and other carbohydrates can also be used - for example, in the lab sucrose (sugar) is normally used to produce ethanol. Industrially, this wouldn't make sense. It would be silly to refine sugar if all you were going to use it for was fermentation. There is no reason why you shouldn't start from the original sugar cane, though.

The first step is to break complex carbohydrates into simpler ones.

For example, if you were starting from starch in grains like wheat or barley, the grain is heated with hot water to extract the starch and then warmed with malt. Malt is germinated barley which contains enzymes which break the starch into a simpler carbohydrate called maltose, $C_{12}H_{22}O_{11}$.

Maltose has the same molecular formula as sucrose but contains two glucose units joined together, whereas sucrose contains one glucose and one fructose unit.

Yeast is then added and the mixture is kept warm (say 35°C) for perhaps several days until fermentation is complete. Air is kept out of the mixture to prevent oxidation of the ethanol produced to ethanoic acid (vinegar).

Enzymes in the yeast first convert carbohydrates like maltose or sucrose into even simpler ones like glucose and fructose, both $C_6H_{12}O_6$, and then convert these in turn into ethanol and carbon dioxide.

You can show these changes as simple chemical equations, but the biochemistry of the reactions is much, much more complicated than this suggests.

$$C_{12}H_{22}O_{11} + H_2O \longrightarrow 2C_6H_{12}O_6$$
$$C_6H_{12}O_6 \longrightarrow 2CH_3CH_2OH + 2CO_2$$

Yeast is killed by ethanol concentrations in excess of about 15%, and that limits the purity of the ethanol that can be produced. The ethanol is separated from the mixture by fractional distillation to give 96% pure ethanol.

For theoretical reasons, it is impossible to remove the last 4% of water by fractional distillation.

A Comparison of Fermentation with the Direct Hydration of Ethene

	Fermentation	*Hydration of ethene*
Type of process	A batch process. Everything is put into a container and then left until fermentation is complete. That batch is then cleared out and a new reaction set up. This is inefficient.	A continuous flow process. A stream of reactants is passed continuously over a catalyst. This is a more efficient way of doing things.
Rate of reaction	Very slow.	Very rapid.
Quality of product	Produces very impure ethanol which needs further processing	Produces much purer ethanol.
Reaction conditions	Uses gentle temperatures and atmospheric pressure.	Uses high temperatures and pressures, needing lots of energy input.
Use of resources	Uses renewable resources based on plant material.	Uses finite resources based on crude oil.

PHYSICAL PROPERTIES

Boiling Points

Lets consider some simple primary alcohols with up to 4 carbon atoms.

They are:

CH_3OH	CH_3CH_2OH	$CH_3CH_2CH_2OH$	$CH_3CH_2CH_2CH_2OH$
methanol	ethanol	propan-1-ol	butan-1-ol

and compare them with the equivalent alkane (methane to butane) with the same number of carbon atoms.

- The boiling point of an alcohol is always much higher than that of the alkane with the same number of carbon atoms.
- The boiling points of the alcohols increase as the number of carbon atoms increases.

Polyhydric alcohols have higher boiling points when compared to similar sized monohydric alcohols. This is easily explained since we have more than one Hydroxyl group in the molecule there would be extensively higher hydrogen bonding between the alcohol molecules.

Hydrogen Bonding

Hydrogen bonding occurs between molecules where you have a hydrogen atom attached to one of the very electronegative elements - fluorine, oxygen or nitrogen.

In the case of alcohols, there are hydrogen bonds set up between the slightly positive hydrogen atoms and lone pairs on oxygens in other molecules.

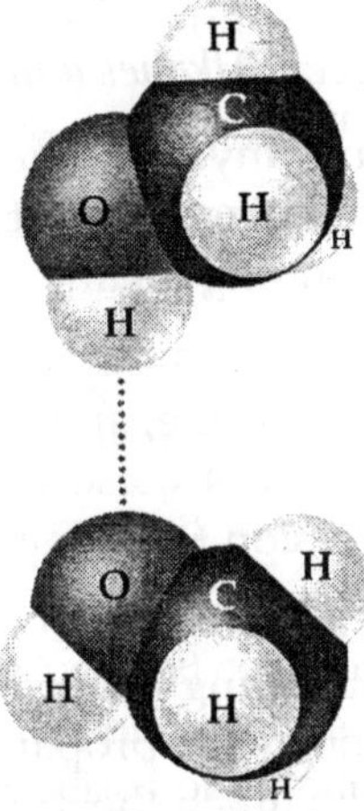

Fig. Hydrogen bonding in methanol

The hydrogen atoms are slightly positive because the bonding electrons are pulled away from them towards the very electronegative oxygen atoms.

In alkanes, the only intermolecular forces are Van der Waals dispersion forces. Hydrogen bonds are much stronger than these and therefore it takes more energy to separate alcohol molecules than it does to separate alkane molecules.

That's the main reason that the boiling points are higher.

The Effect of Van der Waals Forces

On the Boiling Points of the Alcohols

Hydrogen bonding isn't the only intermolecular force in alcohols. There are also Van der Waals dispersion forces and dipole-dipole interactions.

The hydrogen bonding and the dipole-dipole interactions will be much the same for all the alcohols, but the dispersion forces will increase as the alcohols get bigger.

These attractions get stronger as the molecules get longer and have more electrons. That increases the sizes of the temporary dipoles that are set up.

This is why the boiling points increase as the number of carbon atoms in the chains increases. It takes more energy to overcome the dispersion forces, and so the boiling points rise.

On the Comparison between Alkanes and Alcohols

Even if there wasn't any hydrogen bonding or dipole-dipole interactions, the boiling point of the alcohol would be higher than the corresponding alkane with the same number of carbon atoms.

Ethanol is a longer molecule, and the oxygen brings with it an extra 8 electrons. Both of these will increase the size of the Van der Waals dispersion forces and so the boiling point.

If you were doing a really fair comparison to show the effect of the hydrogen bonding on boiling point it would be better to compare ethanol with propane rather than ethane. The length would then be much the same, and the number of electrons is exactly the same.

Solubility of Alcohols in Water

The small alcohols are completely soluble in water. Whatever proportions you mix them in, you will get a single solution.

However, solubility falls as the length of the hydrocarbon chain in the alcohol increases. Once you get to four carbons and beyond, the fall in solubility is noticeable, and you may well end up with two layers in your test tube.

The Solubility of the Small Alcohols in Water

Consider ethanol as a typical small alcohol. In both pure water and pure ethanol the main intermolecular attractions are hydrogen bonds.

In order to mix the two, you would have to break the hydrogen bonds between the water molecules and the hydrogen bonds between the ethanol molecules. It needs energy to do both of these things.

However, when the molecules are mixed, new hydrogen bonds are made between water molecules and ethanol molecules.

The energy released when these new hydrogen bonds are made more or less compensates for that needed to break the original ones.

In addition, there is an increase in the disorder of the system - an increase in entropy. That is another factor in deciding whether things happen or not.

The Lower Solubility of bigger Alcohols

Imagine what happens when you have got, say, 5 carbon atoms in each alcohol molecule.

The hydrocarbon chains are forcing their way between water molecules and so breaking hydrogen bonds between those water molecules.

The -OH end of the alcohol molecules can form new hydrogen bonds with water molecules, but the hydrocarbon "tail" doesn't form hydrogen bonds

That means that quite a lot of the original hydrogen bonds being broken aren't replaced by new ones.

All you get in place of those original hydrogen bonds are Van der Waals dispersion forces between the water and the hydrocarbon "tails". These attractions are much weaker. That means that you don't get enough energy back to compensate for the hydrogen bonds being broken. Even allowing for the increase in disorder, the process becomes less feasible.

As the length of the alcohol increases, this situation just gets worse, and so the solubility falls.

The water solubility of alcohols and ethers are much more similar. Low molecular weight alcohols and ethers are water solubility such as Methanol, Ethanol, and Dimethyl ether. But four carbon alcohols and higher and three carbon ethers or higher have a much lower water solubility. This can be explained by considering the way that water molecules can disperse solute molecules into a solution. The polar water molecules are attracted to the Hydroxyl group by hydrogen bonding that occurs between the hydrogens of the water molecules and the Oxygen of the alcohol or ether. With smaller molecules water can more easily surround the ether or alcohol molecules effectively separating them from one another. This is called solvation and each ether or alcohol molecule is encapsulated in a water "cage". As the hydrocarbon portions of the alcohol or ether become more extensive as they would be in a higher molecular weight alcohol or ether, it takes far more water molecules to achieve the same goal (solvation) if it is possible at all. The larger the solute molecules the less likely that will occur. The water molecules have a greater molecular surface to surround which makes it less likely for solvation.

Increased branching will increase water solubility. This can be explained by the fact that increased branching around the caribinol carbon will mean that water molecules will have a smaller molecular surface area to surround in trying to encapsulate the organic molecule. For example, the water solubility for the same four alcohols mentioned above are 8.3 g/100 ml water, 10.0 grams/100 ml water, 26.0 grams / 100 ml water, and infinite solubility in the case of the 2-methyl-2-propanol(most highly branched of the four).

The water solubilities of polyhydric alcohols are higher when compared to similar sized monohydric alcohols since there are more Hydroxyl groups for water molecules to be attracted to, solvation can more easily be attained.

ACID AND BASE PROPERTIES

Alcohols are capable of losing the Hydrogen atom and thereby acting as an acid, but this is difficult to do. Alcohols tend to be slightly less acidic (pKa = 15) compared to water (pKa = 14). The higher the pKa value the lower is the acid strength. On the other hand the lone pairs found on the Oxygen atom of the Hydroxyl groups make them excellent bases. Most of the chemical activity of alcohols has to do with their basic properties instead of their relatively weak acidic properties.

REACTIONS

With Sodium

Under this heading we will describe the reaction between alcohols and metallic sodium, and takes a very brief look at the properties of the alkoxide which is formed. We will look at the reaction between sodium and ethanol as being typical, but you could substitute any other alcohol you wanted to - the reaction would be the same.

The Reaction between Sodium and Ethanol

Details of the reaction

If a small piece of sodium is dropped into some ethanol, it reacts steadily to give off bubbles of hydrogen gas and leaves a colourless solution of sodium ethoxide, CH3CH2ONa. Sodium ethoxide is known as an *alkoxide*.

If the solution is evaporated carefully to dryness, the sodium ethoxide is left as a white solid.

Although at first sight you might think this was something new and complicated, in fact it is exactly the same (apart from being a more gentle reaction) as the reaction between sodium and water.

Compare the two:

$$2CH_3CH_2\text{-}OH + 2Na \longrightarrow 2CH_3CH_2\text{-}ONa + H_2$$

$$2H\text{-}OH + 2Na \longrightarrow 2H\text{-}ONa + H_2$$

We normally, of course, write the sodium hydroxide formed as NaOH rather than HONa - but that's the only difference.

Sodium ethoxide is just like sodium hydroxide, except that the hydrogen has been replaced by an ethyl group. Sodium hydroxide contains OH- ions; sodium ethoxide contains CH_3CH_2O- ions.

Using the Reaction

There are two simple uses for this reaction:

To Dispose of small Amounts of Sodium Safely

If you spill some sodium on the bench, or have a small amount left over from a reaction, you can't just chuck it in the sink. It tends to react explosively with the water - and comes flying back out at you again!

It reacts much more gently with ethanol. Ethanol is therefore used to dissolve small quantities of waste sodium. The solution formed can be washed away without problems (provided you remember that sodium ethoxide is strongly alkaline - see below).

To Test for the -OH Group in Alcohols

Because of the dangers involved in handling sodium, this is not the best test for an alcohol at this level. Because sodium reacts violently with acids to produce a salt and hydrogen, you would first have to be sure that the liquid you were testing was neutral.

You would also have to be confident that there was no trace of water present because sodium reacts with the -OH group in water even better than with the one in an alcohol.

With those provisos, if you add a tiny piece of sodium to a neutral liquid free of water and get bubbles of hydrogen produced, then the liquid is an alcohol.

Some Simple Reactions of Alkoxide Ions

Once again we will take the ethoxide ions in sodium ethoxide as typical. Essentially, ethoxide (and other alkoxide) ions behave like hydroxide ions.

Ethoxide Ions are Strongly Basic

If you add water to sodium ethoxide, it dissolves to give a colourless solution with a high pH - typically pH 14. The solution is strongly alkaline.

The reason is that the ethoxide ions remove hydrogen ions from water molecules to produce hydroxide ions. It is these which produce the high pH.

$$CH_3CH_2O^- + H_2O \longrightarrow CH_3CH_2OH + OH^-.$$

Ethoxide Ions are Good Nucleophiles

A nucleophile is something which carries a negative or partial negative charge which it uses to attack positive centres in other molecules or ions.

Hydroxide ions are good nucleophiles, and you may have come across the reaction between a halogenoalkane (also called a haloalkane or alkyl halide) and sodium hydroxide solution. The hydroxide ions replace the halogen atom.

$$CH_3CH_2CH_2Br + OH^- \longrightarrow CH_3CH_2CH_2OH + Br$$

In this case, an alcohol is formed.

The ethoxide ion behaves in exactly the same way. If you knew the mechanism for the hydroxide ion reaction you could work out exactly what happens in the reaction between a halogenoalkane and ethoxide ion.

Compare this equation with the last one.

$$CH_3CH_2CH_2Br + CH_3CH_2O^- \longrightarrow CH_3CH_2CH_2OCH_2CH_3 + Br$$

$$CH_3CH_2CH_2Br + OH^- \longrightarrow CH_3CH_2CH_2OH + Br$$

The only difference is that where there was a hydrogen atom at the right-hand end of the product molecule, you now have an alkyl group.

Two alkyl (or other hydrocarbon) groups bridged by an oxygen atom is called an *ether*. This particular one is 1-ethoxypropane or ethyl propyl ether. This reaction is a good way of making ethers in the lab.

REPLACING THE -OH GROUP BY A HALOGEN

Reactions Involving Hydrogen Halides

The general reaction looks like this:

$$ROH + HX \longrightarrow RX + H_2O$$

Reaction with Hydrogen Chloride

Tertiary alcohols react reasonably rapidly with concentrated hydrochloric acid, but for primary or secondary alcohols the reaction rates are too slow for the reaction to be of much importance.

A tertiary alcohol reacts if it is shaken with concentrated hydrochloric acid at room temperature. A tertiary halogenoalkane (haloalkane or alkyl halide) is formed

$$CH_3\text{-}C(CH_3)_2\text{-}OH + HCl \longrightarrow CH_3\text{-}C(CH_3)_2\text{-}Cl + H_2O$$

Replacing -OH by bromine

Rather than using hydrobromic acid, you usually treat the alcohol with a mixture of sodium or potassium bromide and concentrated sulphuric acid. This produces hydrogen bromide which reacts with the alcohol. The mixture is warmed to distil off the bromoalkane.

$$CH_3\text{-}CH_2\text{-}OH + HBr \longrightarrow CH_3\text{-}CH_2\text{-}Br + H_2O$$

Replacing -OH by iodine

In this case the alcohol is reacted with a mixture of sodium or potassium iodide and concentrated phosphoric(V) acid, H_3PO_4, and the iodoalkane is distilled off. The mixture of the iodide and phosphoric(V) acid produces hydrogen iodide which reacts with the alcohol.

$$CH_3\text{-}CH_2\text{-}OH + HI \longrightarrow CH_3\text{-}CH_2\text{-}I + H_2O$$

Phosphoric(V) acid is used instead of concentrated sulphuric acid because sulphuric acid oxidizes iodide ions to iodine and produces hardly any hydrogen iodide.

A similar thing happens to some extent with bromide ions in the preparation of bromoalkanes, but not enough to get in the way of the main reaction. There is no reason why you couldn't use phosphoric(V) acid in the bromide case instead of sulphuric acid if you wanted to.

REACTING ALCOHOLS WITH PHOSPHORUS HALIDES

Reaction with Phosphorus(III) Chloride, PCl_3

Alcohols react with liquid phosphorus(III) chloride (also called phosphorus trichloride) to make chloroalkanes.

$$3CH_3\text{-}CH_2\text{-}CH_2\text{-}OH + PCl_3 \longrightarrow 3CH_3\text{-}CH_2\text{-}CH_2\text{-}Cl + H_3PO_3$$

Reaction with Phosphorus(V) Chloride, PCl_5

Solid phosphorus(V) chloride (phosphorus pentachloride) reacts violently with alcohols at room temperature, producing clouds of hydrogen chloride gas. It isn't a good choice as a way of making chloroalkanes, although it is used as a *test for -OH groups* in organic chemistry.

To show that a substance was an alcohol, you would first have to eliminate all the other things which also react with phosphorus(V) chloride. For example, carboxylic acids (containing the -COOH group) react with it (because of the -OH in -COOH), and so does water (H-OH).

If you have a neutral liquid not contaminated with water, and get clouds of hydrogen chloride when you add phosphorus(V) chloride, then you have an alcohol group present.

$$CH_3\text{-}CH_2\text{-}CH_2\text{-}OH + PCl_5 \longrightarrow CH_3\text{-}CH_2\text{-}CH_2\text{-}Cl + POCl_3 + HCl$$

There are also side reactions involving the $POCl_3$ reacting with the alcohol.

Other Reactions Involving Phosphorus Halides

Instead of using phosphorus(III) bromide or iodide, the alcohol is usually heated under reflux with a mixture of red phosphorus and either bromine or iodine.

The phosphorus first reacts with the bromine or iodine to give the phosphorus(III) halide.

$$2P + 3Br_2 \longrightarrow 2PBr_3$$
$$2P + 3I_2 \longrightarrow 2PI_3$$

These then react with the alcohol to give the corresponding halogenoalkane which can be distilled off.

$$3CH_3\text{-}CH_2\text{-}OH + PBr_3 \longrightarrow 3CH_3\text{-}CH_2\text{-}Br + H_3PO_3$$

$$3CH_3\text{-}CH_2\text{-}OH + PI_3 \longrightarrow 3CH_3\text{-}CH_2\text{-}I + H_3PO_3$$

REACTING ALCOHOLS WITH SULPHUR DICHLORIDE OXIDE (THIONYL CHLORIDE)

Sulphur dichloride oxide (thionyl chloride) has the formula $SOCl_2$.

Traditionally, the formula is written as shown, despite the fact that the modern *name* writes the chlorine before the oxygen (alphabetical order).

The sulphur dichloride oxide reacts with alcohols at room temperature to produce a chloroalkane. Sulphur dioxide and hydrogen chloride are given off. Care would have to be taken because both of these are poisonous.

$$CH_3\text{-}CH_2\text{-}CH_2\text{-}OH + SOCl_2 \longrightarrow CH_3\text{-}CH_2\text{-}CH_2\text{-}Cl + SO_2$$

The big adVantage that this reaction has over the use of either of the phosphorus chlorides is that the two other products of the reaction (sulphur dioxide and HCl) are both gases. That means that they separate themselves from the reaction mixture.

OXIDATION

Here we are going to discuss the oxidation of alcohols using acidified sodium or potassium dichromate(VI) solution. This reaction is used to make aldehydes, ketones and carboxylic acids, and as a way of distinguishing between primary, secondary and tertiary alcohols.

Oxidizing the Different Types of Alcohols

The oxidizing agent used in these reactions is normally a solution of sodium or potassium dichromate(VI) acidified with dilute sulphuric acid. If oxidation occurs, the orange solution containing the dichromate(VI) ions is reduced to a green solution containing chromium(III) ions.

The electron-half-equation for this reaction is

$$Cr_2O_7^{2-} + 14H^+ + 6e^- \longrightarrow 2Cr^{3+} + 7H_2O$$

Primary Alcohols

Primary alcohols can be oxidized to either aldehydes or carboxylic acids depending on the reaction conditions. In the case of the formation of carboxylic acids, the alcohol is first oxidized to an aldehyde which is then oxidized further to the acid.

Partial Oxidation to Aldehydes

You get an aldehyde if you use an excess of the alcohol, and distil off the aldehyde as soon as it forms.

The excess of the alcohol means that there isn't enough oxidising agent present to carry out the second stage. Removing the aldehyde as soon as it is formed means that it doesn't hang around waiting to be oxidized anyway!

If you used ethanol as a typical primary alcohol, you would produce the aldehyde ethanal, CH_3CHO.

The full equation for this reaction is fairly complicated, and you need to understand about electron-half-equations in order to work it out.

$$3CH_3CH_2OH + Cr_2O_7^{2-} + BH^+ \longrightarrow 3CH_3CHO + 2Cr^{3+} + 7H_2O$$

In organic chemistry, simplified versions are often used which concentrate on what is happening to the organic substances. To do that, oxygen from an oxidising agent is represented as [O]. That would produce the much simpler equation:

$$CH_3CH_2OH + [O] \longrightarrow CH_3CHO + H_2O$$

This means "oxygen from an oxidising agent"

It also helps in remembering what happens. You can draw simple structures to show the relationship between the primary alcohol and the aldehyde formed.

$$CH_3\text{-}CH_2\text{-}OH \xrightarrow{[O]} CH_3\text{-}CHO + H_2O$$

Full Oxidation to Carboxylic Acids

You need to use an excess of the oxidising agent and make sure that the aldehyde formed as the half-way product stays in the mixture.

The alcohol is heated under reflux with an excess of the oxidising agent. When the reaction is complete, the carboxylic acid is distilled off.

The full equation for the oxidation of ethanol to ethanoic acid is:

$$3CH_3CH_2OH + 2Cr_2O_7^{2-} + 16H^+ \longrightarrow 3CH_3COOH + 4Cr^{3+} + 11H_2O$$

The more usual simplified version looks like this:

$$CH_3CH_2OH + 2[O] \longrightarrow CH_3COOH + H_2O$$

Alternatively, you could write separate equations for the two stages of the reaction - the formation of ethanal and then its subsequent oxidation.

$$CH_3CH_2OH + [O] \longrightarrow CH_3CHO + H_2O$$

$$CH_3CHO + [O] \longrightarrow CH_3COOH$$

This is what is happening in the second stage:

$$CH_3\text{-}C(=O)\text{-}H + [O] \longrightarrow CH_3\text{-}C(=O)\text{-}OH$$

Secondary Alcohols

Secondary alcohols are oxidized to ketones - and that's it. For example, if you heat the secondary alcohol propan-2-ol with sodium or potassium dichromate(VI) solution acidified with dilute sulphuric acid, you get propanone formed.

Playing around with the reaction conditions makes no difference whatsoever to the product.

Using the simple version of the equation and showing the relationship between the structures:

$$CH_3\text{-}CH(CH_3)\text{-}OH + [O] \longrightarrow CH_3\text{-}C(=O)\text{-}CH_3 + H_2O$$

propan-2-ol → propanone

If you look back at the second stage of the primary alcohol reaction, you will see that an oxygen "slotted in" between the carbon and the hydrogen in the aldehyde group to produce the carboxylic acid. In this case, there is no such hydrogen - and the reaction has nowhere further to go.

Tertiary Alcohols

Tertiary alcohols aren't oxidized by acidified sodium or potassium dichromate(VI) solution. There is no reaction whatsoever.

If you look at what is happening with primary and secondary alcohols, you will see that the oxidising agent is removing the hydrogen from the -OH group, and a hydrogen from the carbon atom attached to the -OH. Tertiary alcohols don't have a hydrogen atom attached to that carbon.

You need to be able to remove those two particular hydrogen atoms in order to set up the carbon-oxygen double bond.

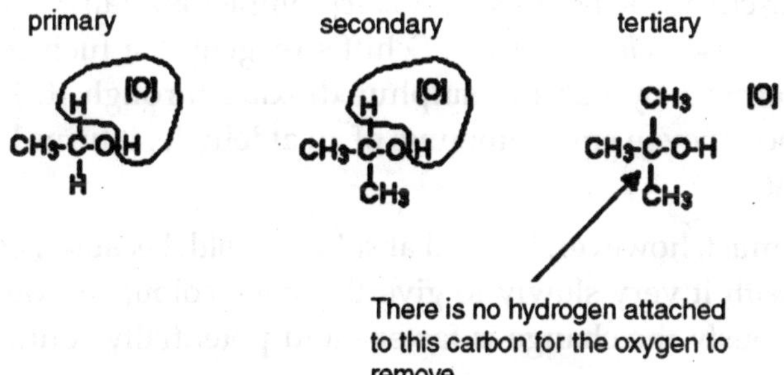

Using these Rreactions as a Test for the Different Types of Alcohol

First you have to be sure that you have actually got an alcohol by testing for the -OH group. You would need to show that it was a neutral liquid, free of water and that it reacted with solid phosphorus(V) chloride to produce a burst of acidic steamy hydrogen chloride fumes. You would then add a few drops of the alcohol to a test tube containing potassium dichromate(VI) solution acidified with dilute sulphuric acid. The tube would be warmed in a hot water bath.

Results for the Various Kinds of Alcohol

Picking out the Tertiary Alcohol

In the case of a primary or secondary alcohol, the orange solution turns green on heating. With a tertiary alcohol there is no colour change.

Distinguishing between the Primary and Secondary Alcohols

You need to produce enough of the aldehyde (from oxidation of a primary alcohol) or ketone (from a secondary alcohol) to be able to test them. There are various things which aldehydes do which ketones don't. These include the reactions with Tollens' reagent, Fehling's solution and Benedict's solution, and are covered on a separate page. These tests can be a bit of a bother to carry out and the results aren't always as clear-cut as the books say. A *much* simpler but fairly reliable test is to use *Schiff's reagent.* Schiff's reagent is a fuchsin dye decolourised by passing sulphur dioxide through it. In the presence of even small amounts of an aldehyde, it turns bright magenta.

It must, however, be used absolutely cold, because ketones react with it very slowly to give the same colour. If you heat it, obviously the change is faster - and potentially confusing.

While you are warming the reaction mixture in the hot water bath, you can pass any vapours produced through some Schiff's reagent.

- If the Schiff's reagent quickly becomes magenta, then you are producing an aldehyde from a primary alcohol.
- If there is no colour change in the Schiff's reagent, or only a trace of pink colour within a minute or so, then

you aren't producing an aldehyde, and so haven't got a primary alcohol.

Because of the colour change to the acidified potassium dichromate(VI) solution, you must therefore have a secondary alcohol.

You should check the result as soon as the potassium dichromate(VI) solution turns green - if you leave it too long, the Schiff's reagent might start to change colour in the secondary alcohol case as well.

THE TRIIODOMETHANE (IODOFORM) REACTION

The triiodomethane (iodoform) reaction can be used to identify the presence of a CH3CH(OH) group in alcohols.

Doing the Triiodomethane (iodoform) Reaction

There are two apparently quite different mixtures of reagents that can be used to do this reaction. They are, in fact, chemically equivalent.

Using Iodine and Sodium Hydroxide Solution

This is chemically the more obvious method.

Iodine solution is added to a small amount of an alcohol, followed by just enough sodium hydroxide solution to remove the colour of the iodine. If nothing happens in the cold, it may be necessary to warm the mixture very gently.

A positive result is the appearance of a very pale yellow precipitate of triiodomethane (previously known as iodoform) - CHI_3.

Apart from its colour, this can be recognized by its faintly "medical" smell. It is used as an antiseptic on the sort of sticky plasters you put on minor cuts, for example.

Using Potassium Iodide and Sodium Chlorate(I) Solutions

Sodium chlorate(I) is also known as sodium hypochlorite.

Potassium iodide solution is added to a small amount of an alcohol, followed by sodium chlorate(I) solution. Again, if no precipitate is formed in the cold, it may be necessary to warm the mixture very gently.

The positive result is the same pale yellow precipitate as before.

CHEMISTRY OF THE TRIIODOMETHANE (IODOFORM) REACTION

What the Triiodomethane (iodoform) Reaction Shows

A positive result - the pale yellow precipitate of triiodomethane (iodoform) - is given by an alcohol containing the grouping:

$$CH_3-\overset{\displaystyle H}{\underset{\displaystyle R}{\overset{|}{\underset{|}{C}}}}-O\text{-}H$$

"R" can be a hydrogen atom or a hydrocarbon group (for example, an alkyl group).

If "R" is hydrogen, then you have the primary alcohol ethanol, CH_3CH_2OH.

- Ethanol is the *only* primary alcohol to give the triiodomethane (iodoform) reaction.
- If "R"is a hydrocarbon group, then you have a secondary alcohol. Lots of secondary alcohols give this reaction, but those that do all have a methyl group attached to the carbon with the -OH group.
- *No* tertiary alcohols can contain this group because no tertiary alcohols can have a hydrogen atom attached to the carbon with the -OH group. No tertiary alcohols give the triiodomethane (iodoform) reaction.

Summary of the Reactions During the Triiodomethane (iodoform) Reaction

We will take the reagents as being iodine and sodium hydroxide solution.

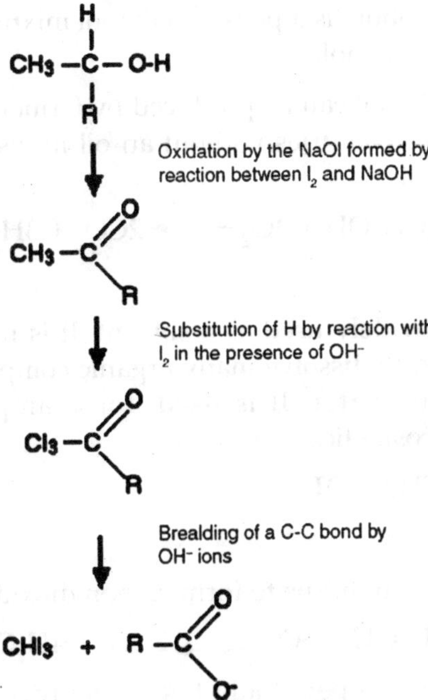

This is being given as a flow scheme rather than full equations.

USES OF ALCOHOLS

USES OF ETHANOL

Drinks

The "alcohol" in alcoholic drinks is simply ethanol.

Industrial Methylated Spirits (meths)

Ethanol is usually sold as industrial methylated spirits which is ethanol with a small quantity of methanol added and possibly some colour. Methanol is poisonous, and so the industrial methylated spirits is unfit to drink. This avoids the high taxes which are levied on alcoholic drinks.

As a Fuel

Ethanol burns to give carbon dioxide and water and can be used as a fuel in its own right, or in mixtures with petrol (gasoline). "Gasohol" is a petrol / ethanol mixture containing about 10 - 20% ethanol.

Because ethanol can be produced by fermentation, this is a useful way for countries without an oil industry to reduce imports of petrol.

$$CH_3CH_2OH + 3O_2 \longrightarrow 2CO_2 + 3H_2O$$

As a Solvent

Ethanol is widely used as a solvent. It is relatively safe, and can be used to dissolve many organic compounds which are insoluble in water. It is used, for example, in many perfumes and cosmetics.

USES OF METHANOL

As a Fuel

Methanol again burns to form carbon dioxide and water.

$$2CH_3OH + 3O_2 \longrightarrow 2CO_2 + 4H_2O$$

It can be used a petrol additive to improve combustion, or work is currently being done on its use as a fuel in its own right.

As an Industrial Feedstock

Most methanol is used to make other things - for example, methanal (formaldehyde), ethanoic acid, and methyl esters of various acids. In most cases, these are in turn converted into further products.

Chapter 10

Ether

INTRODUCTION

Ether is the general name for a class of chemical compounds which contain an ether group — an oxygen atom connected to two (substituted) alkyl groups. A typical example is the solvent and anesthetic diethyl ether, commonly referred to simply as "ether", (ethoxyethane, CH_3-CH_2-O-CH_2-CH_3).

NOMENCLATURE

In the IUPAC nomenclature system, ethers are named using the general formula *"alkoxyalkane"*, for example CH_3-CH_2-O-CH_3 is methoxyethane. If the ether is part of a more complex molecule, it is described as an alkoxy substituent, so -OCH_3 would be considered a *"methoxy-"* group. The nomenclature of describing the two alkyl groups and appending *"ether"*, e.g. *"ethyl methyl ether"* in the example above, is a trivial usage.

Ethers (R-O-R) consist of an oxygen atom between the two attached carbon chains. The shorter of the two chains becomes the first part of the name with the -ane suffix changed to -oxy, and the longer alkane chain become the suffix of the name of the ether. Thus CH_3OCH_3 is methoxymethane, and $CH_3OCH_2CH_3$ is methoxyethane (*not* ethoxymethane). If the oxygen is not attached to the end of the main alkane chain, then the whole shorter alkyl-plus-ether group is treated as a side-chain and prefixed with its bonding position on the main chain. Thus $CH_3OCH(CH_3)_2$ is 2-methoxypropane.

Primary, Secondary, and Tertiary Ethers

The terms *"primary ether"*, *"secondary ether"*, and *"tertiary ether"* are occasionally used and refer to the carbon atom next to the ether oxygen. In a *primary ether* this carbon is connected to only one other carbon as in diethyl ether CH_3-CH_2-O-CH_2-CH_3. An example of a *secondary ether* is diisopropyl ether $(CH_3)_2$CH-O-CH$(CH_3)_2$ and that of a *tertiary ether* is di-tert-butyl ether $(CH_3)_3$C-O-C$(CH_3)_3$.

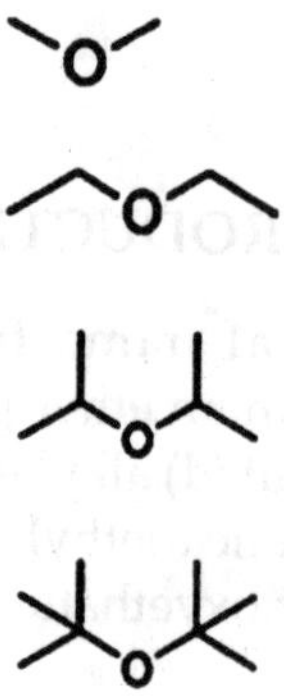

Dimethyl ether, a *primary*, a *secondary*, and a *tertiary ether.*

PREPARATION

Ethers can be prepared in the laboratory in several different ways.

DEHYDRATION OF ALCOHOLS

R-OH + R-OH $\rightarrow$ R-O-R + H_2O

This direct reaction requires drastic conditions (heat and an acid catalyst) and is usually not applicable. Such conditions can destroy the delicate structures of some functional groups. There exist several milder methods to produce ethers.

NUCLEOPHILIC DISPLACEMENT OF ALKYL HALIDES BY ALKOXIDES

$R\text{-}O^- + R\text{-}X \rightarrow R\text{-}O\text{-}R + X^-$

This reaction is called the Williamson ether synthesis. It involves treatment of a parent alcohol with a strong base to form the alkoxide anion followed by addition of an appropriate aliphatic compound bearing a suitable leaving group (R-X). Suitable leaving groups (X) include iodide, bromide, or sulfonates. This method does not work if R is aromatic like in bromobenzene. Likewise, this method only gives the best yields for primary carbons, as secondary carbons will undergo E_2 elimination on exposure to the basic alkoxide anion used in the reaction. Aryl ethers can be prepared in the Ullmann condensation.

ELECTROPHILIC ADDITION OF ALCOHOLS TO ALKENES

$$R_2C{=}CR_2 + R\text{-}OH \rightarrow R_2CH\text{-}C(\text{-}O\text{-}R)\text{-}R_2$$

Acid catalysis is required for this reaction. Tetrahydropyranyl ethers are used as protective groups for alcohols.

Cyclic ethers which are also known as epoxides can be prepared:

- By the oxidation of alkenes with a peroxyacid.
- By the base intramolecular nuclephilic substitution of a halohydrin.

PHYSICAL PROPERTIES

Ether molecules cannot form hydrogen bonds among each other, resulting in a relatively low boiling point comparable to that of the analogous alkanes. Ethers are more hydrophobic than esters or amides of comparable structure.

Ethers can act as Lewis bases. For instance, diethyl ether forms a complex with boron compounds, such as boron trifluoride diethyl etherate $F_3B{:}O(CH_2CH_3)_2$. Ethers also coordinate to magnesium in Grignard reagents.

REACTIONS

Ethers in general are of very low chemical reactivity. Organic reactions are:

Hydrolysis

Ethers are hydrolyzed only under drastic conditions like heating with boron tribromide or boiling in hydrobromic acid.

Lower mineral acids containing a halogen, such as hydrochloric acid will cleave ethers, but very slowly. Hydrobromic acid and hydroiodic acid are the only two that do so at an appreciable rate. Certain aryl ethers can be cleaved by aluminium chloride.

Nucleophilic Displacement

Epoxides, or cyclic ethers in three-membered rings, are highly susceptible to nucleophilic attack and are reactive in this fashion.

The reason that epoxides or epoxyethane is so reactive is that bonding pairs in the ring of atoms in the molecule are forced very close together. The bond angles are about 60° rather than about 109.5° when carbon atoms normally form single bonds.

The overlap between the atomic orbitals in forming the carbon-carbon and carbon-oxygen bonds is less good than it is normally, and there is considerable repulsion between the bonding pairs. The system becomes more stable if the ring is broken.

When epoxyethane reacts a carbon-oxygen bond is always broken and the ring opens up.

Peroxide Formation

Primary and secondary ethers with a CH group next to the ether oxygen easily form highly explosive organic peroxides (e.g. diethyl ether peroxide) in the presence of oxygen, light, and metal and aldehyde impurities. For this reason ethers like diethyl ether and THF are usually avoided as solvents in industrial processes.

Chapter 11

Aldehydes and Ketones

INTRODUCTION

Aldehydes and ketones are simple compounds which contain a *carbonyl group* - a carbon-oxygen double bond. They are simple in the sense that they don't have other reactive groups like -OH or -Cl attached directly to the carbon atom in the carbonyl group - as you might find, for example, in carboxylic acids containing -COOH.

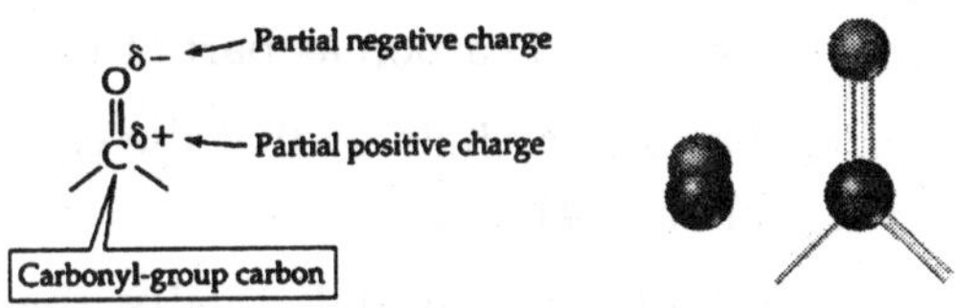

Fig. Carbonyl group

Examples of Aldehydes

In aldehydes, ***the carbonyl group has a hydrogen atom attached to it*** together with either

- a second hydrogen atom
- or, more commonly, a hydrocarbon group which might be an alkyl group or one containing a benzene ring.

Examples of aldehydes are:

$H-C(=O)-H$	$CH_3-C(=O)-H$	$CH_3-CH_2-C(=O)-H$	$CH_3-CH_2-CH(CH_3)-C(=O)-H$
methanal	ethanal	propanal	2-methylbutanal

Notice that these all have exactly the same end to the molecule. All that differs is the complexity of the other group attached.

Examples of ketones

Notice that ketones *never* have a hydrogen atom attached to the carbonyl group.

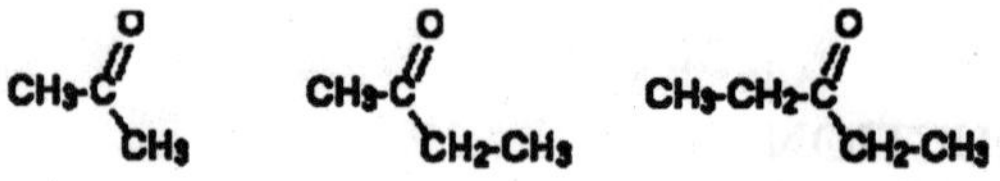

propanone butanone pentan-3-one

IUPAC NOMENCLATURE OF ALDEHYDES

Aldehydes are named by using the following rules:

1. Identify the longest continuous chain of carbons with the acyl or carbonyl carbon as part of the chain.
2. Number the carbon chain so that the carbonyl (acyl) carbon is always #1.
3. Locate and identify alphabetically the branched groups by prefixing the carbon number it is attached to. If more than one of the same type of branched group is involved use the Greek prefixes di for 2, tri for three, etc.
4. After identifying the name, number and location of each branched group, use the alkane name corresponding to the number of carbons in the continuous chain
5. Drop the "e" and add the characteristic IUPAC ending for all aldehydes, "al"
6. Alkenals involving Pi bonding will require that the Pi bond is located but the ending will still be "al"

Here are three examples shown in Fig 1 below.

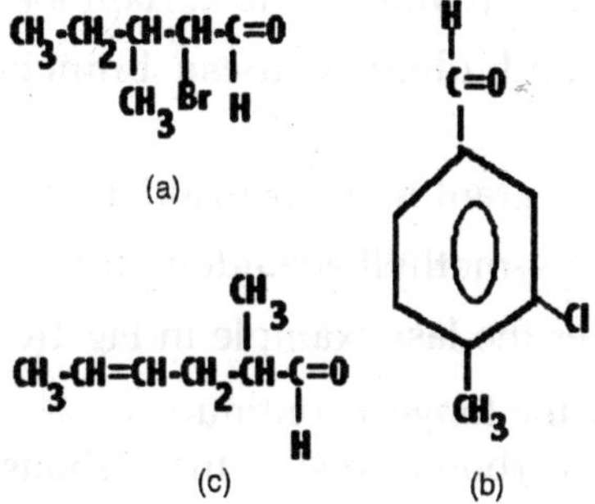

Fig. Structural Examples of Aldehydes

Let us consider the structure in Fig 1(a).

1. We find the longest continuous chain of carbons with the acyl carbon involved is five.
2. Numbering the carbons beginning with the acyl carbon on the extreme right as carbon #1
3. Identifying the branched groups, there is a methyl group on carbon #3 and a Bromine on carbon #2 so we would name and locate them: 2-Bromo-3-methyl
4. use the alkane name corresponding to the number of carbons in the chain (5) which would be pentane
5. drop the "e" and add "al" so the name is: 2-Bromo-3-methylpentanal

Let's consider the structure in Fig 1(b) above.

1. Notice that there is a benzene ring with the characteristic functional group attached to the ring. This would be the parent aromatic aldehyde, benzaldehyde
2. Since we have three substitutions on the Benzene ring we must use numbers and number the ring carbons beginning with the carbon with the aldehyde functional group attached to it as carbon #1. We then proceed to number clockwise.

3. We notice a Chlorine attached to carbon #3 and a methyl group attached to carbon #4.
4. Locate and identify these branches: 3-Chloro-4-methyl
5. Add the parent name benzaldehyde and we have:

   3-Chloro-4-methylbenzaldehyde.

Lets consider the last example in Fig 1(c) above

1. Identify the longest continuous chain of carbons with the acyl carbon as one of the carbons and the double bond must be between two of the carbons in the continuous chain which would be six carbons.
2. Number the carbons in the chain so that the acyl carbon is carbon #1
3. Locate and identify all branches which is only a methyl attached to carbon # 2.

   2-methyl
4. Locate the Pi bond in the chain. It is between the carbon #4 and 5
5. Use the alkene name corresponding to the number of carbons (6) hexene
6. Locate the Pi bond by prefixing the lowest carbon # in which the Pi bond is between. It is between carbon #4 and 5 so it would be

   4-hexene
7. Add to the name and it becomes:

   2-methyl-4-hexene
8. Drop the "e" and add the characteristic "al" and it becomes:

   2-methyl-4-hexenal

IUPAC NOMENCLATURE OF KETONES

Ketones are named following the steps below:

1. Identify the longest continuous chain of carbons with the carbonyl carbon as on of the carbons in the chain.

2. Number the carbons so that the carbonyl carbon has the lowest possible number. If it makes no difference to the carbonyl carbon then number so that the branched groups are attached to the lowest possible carbon numbers
3. Identify the branched groups by naming them in alphabetical order and prefixing the carbon number they are attached to onto the name. If more than one of the same group is present use the Greek prefixes (di for 2 tri for 3, etc)
4. Use the alkane name corresponding to the number of carbons in the chain and prefix the carbon number that the carbonyl carbon has to the front of the alkane name.
5. Drop the "e" in the alkane name and add the characteristic IUPAC ending for ketones which is "one"

Let's try some examples found below in Fig 3.

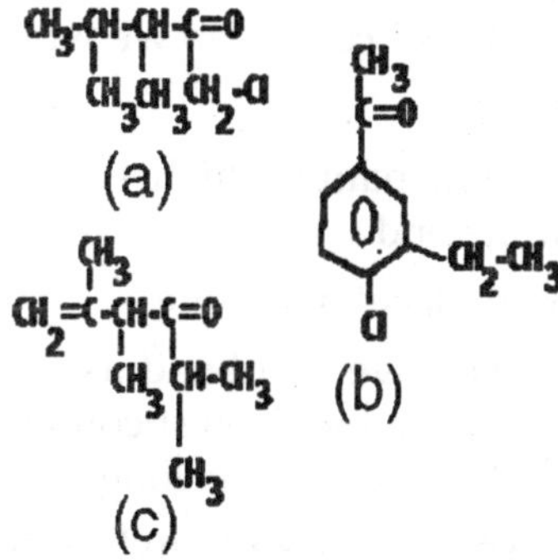

Fig. Keytone practice examples

1. In Fig 3(a), the longest continuous chain with the carbonyl carbon involved is five carbons in length.
2. Numbering so that the carbonyl carbon has the lowest number will mean numbering right to left in the above structure.
3. There are two branched groups that are methyls on the #3 and #4 carbons and a Chlorine on carbon #1.

Naming these alphabetically will give:

1-Chloro-3,4-dimethyl

4. Going back to the chain there are five carbons in the chain so use the alkane name corresponding to the five carbon alkane. The carbonyl group has to be located by prefixing the carbonyl carbon number to the alkane name.

   1-Chloro-3,4-dimethyl-2-pentane

5. Drop the "e" on the alkane name and add the characteristic IUPAC name for a ketone, "one"

   1-Chloro-3,4-dimethyl-2-pentanone

1. In Fig 3(b) we have an aromatic ketone. The parent compound is called acetophenone. There are two groups attached to the ring, an ethyl and a chlorine.
2. Number the carbons in the ring starting with the carbon that has the carbonyl carbon attached directly to the ring carbon.
3. There is an ethyl group on the #3 carbon and a Chlorine on the #4 carbon so we would have:

   4-Chloro-3-ethyl

4. Note the parent name which is acetophenone and attach it on the end

   4-Chloro-3-ethylacetophenone

1. For Fig 3(c) we have a Pi bond as well as a carbonyl carbon. Identifying the longest continuous chain with both involved is six carbons long.
2. Numbering so that the carbonyl carbon is the lowest possible numbered would mean numbering from right to left in the structure above.
3. Identifying the branched groups there are three methyl groups attached to the #2, #4, and #5 carbons in the chain. Naming these gives

   2,4,5-trimethyl

4. locating the Pi bond laces it between the #5 and #6 carbons

5. Using the alkene name for six carbons (hexene) and prefixing the lowest carbon number onto the front of that name gives

   2,4,5-trimethyl-5-hexene

6. Dropping the "e" on alkene and inserting the carbon number (3) that represents the carbonyl carbon replace the "e" with "-3-one" and the name of the compound is

   2,4,5-trimethyl-5-hexen-3-one

BONDING IN CARBONYL COMPOUNDS

Bonding in the Carbonyl Group

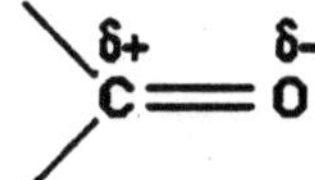

Oxygen is far more electronegative than carbon and so has a strong tendency to pull electrons in a carbon-oxygen bond towards itself. One of the two pairs of electrons that make up a carbon-oxygen double bond is even more easily pulled towards the oxygen. That makes the carbon-oxygen double bond very highly polar.

The Carbonyl Group

The Simple view of the Bonding in Carbon - Oxygen Double Bonds

Where the carbon-oxygen double bond, C=O, occurs in organic compounds it is called a *carbonyl group*. The simplest compound containing this group is methanal.

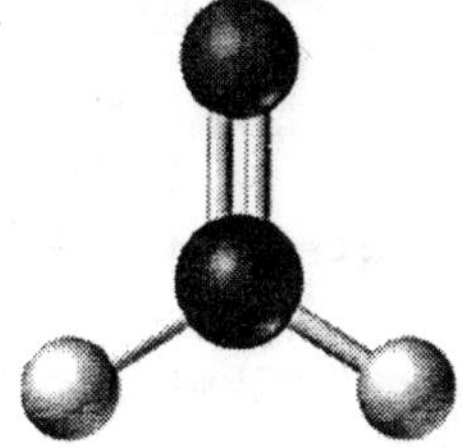

Fig. Methanal or Formaldehyde

We are going to look at the bonding in methanal, but it would equally apply to any other compound containing C=O. The interesting thing is the nature of the carbon-oxygen double bond - not what it's attached to.

An Orbital view of the Bonding in Carbon - Oxygen double Bonds

The Carbon Atom

Just as in ethene or benzene, the carbon atom is joined to three other atoms. The carbon's electrons rearrange themselves, and promotion and hybridisation give sp^2 hybrid orbitals.

Promotion gives:

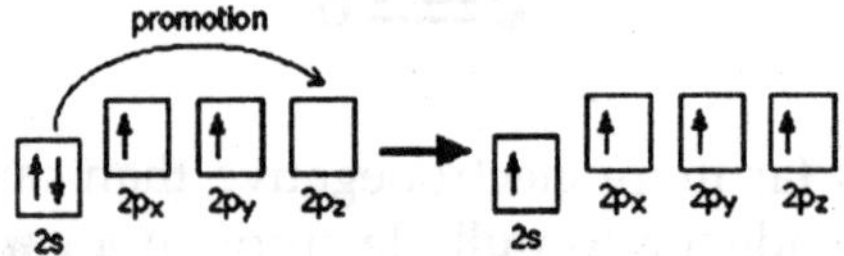

Three sp^2 hybrid orbitals formed arrange themselves as far apart in space as they can - at 120° to each other. The remaining p orbital is at right angles to them.

The Oxygen Atom

Oxygen's electronic structure is $1s^2 2s^2 2p_x^2 2p_y^1 2p_z^1$.

The 1s electrons are too deep inside the atom to be concerned with the bonding and so we'll ignore them from now on. Hybridisation occurs in the oxygen as well. It is easier to see this using "electrons-in-boxes".

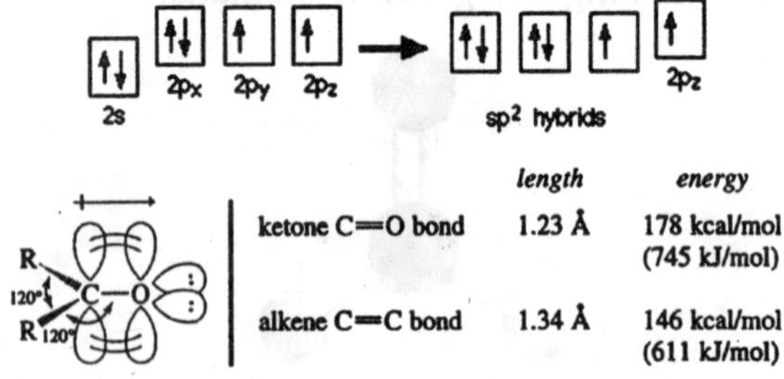

Fig. Bonding in aldehyde and ketone

This distortion in the pi bond causes major differences in the reactions of compounds containing carbon-oxygen double bonds like methanal compared with compounds containing carbon-carbon double bonds like ethene.

PREPARATION

Oxidizing Alcohols to make Aldehydes and Ketones

The oxidizing agent used in these reactions is normally a solution of sodium or potassium dichromate(VI) acidified with dilute sulphuric acid. If oxidation occurs, the orange solution containing the dichromate(VI) ions is reduced to a green solution containing chromium(III) ions.

The net effect is that an oxygen atom from the oxidizing agent removes a hydrogen from the -OH group of the alcohol and one from the carbon to which it is attached.

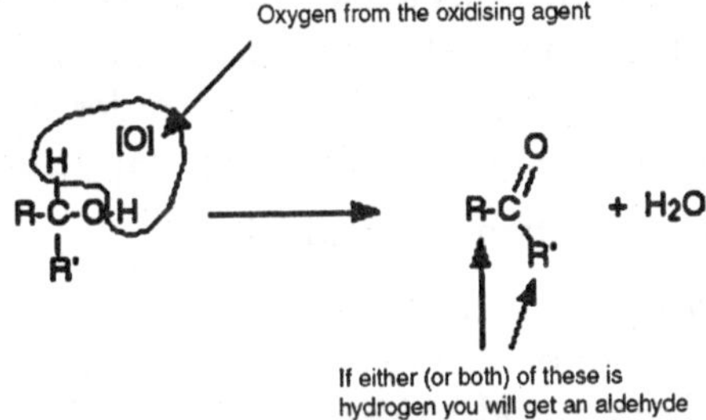

If both R and R' are alkyl groups
you will get a ketone

[O] is often used to represent oxygen coming from an oxidising agent.

R and R′ are alkyl groups or hydrogen. They could also be groups containing a benzene ring, but I'm ignoring these to keep things simple.

If at least one of these groups is a hydrogen atom, then you will get an aldehyde. If they are both alkyl groups then you get a ketone.

If you now think about where they are coming from, you will get an aldehyde if your starting molecule looks like this:

H
R-C-O-H This is a primary alcohol
H

In other words, if you start from a primary alcohol, you will get an aldehyde.

You will get a ketone if your starting molecule looks like this:

```
   H
   |
 R-C-O-H   This is a secondary alcohol
   |
   R'
```

where R and R′ are both alkyl groups.

Secondary alcohols oxidize to give ketones.

Making Aldehydes

Aldehydes are made by oxidising primary alcohols. There is, however, a problem.

The aldehyde produced can be oxidized further to a carboxylic acid by the acidified potassium dichromate(VI) solution used as the oxidising agent. In order to stop at the aldehyde, you have to prevent this from happening.

To stop the oxidation at the aldehyde, you . . .

- use an excess of the alcohol. That means that there isn't enough oxidising agent present to carry out the second stage and oxidize the aldehyde formed to a carboxylic acid.
- distil off the aldehyde as soon as it forms. Removing the aldehyde as soon as it is formed means that it doesn't stay in the mixture to be oxidized further.If you used ethanol as a typical primary alcohol, you would produce the aldehyde ethanal, CH_3CHO.The full equation for this reaction is fairly complicated, and you need to understand about electron-half-equations in order to work it out.

$$3CH_3CH_2OH + Cr_2O_7^{2-} + BH^+ \longrightarrow 3CH_3CHO + 2Cr^{3+} + 7H_2O$$

In organic chemistry, simplified versions are often used which concentrate on what is happening to the organic substances. To do that, oxygen from an oxidising agent is represented as [O]. That would produce the much simpler equation:

$$CH_3CH_2OH + [O] \longrightarrow CH_3CHO + H_2O$$

This means "oxygen from an oxidising agent"

Secondary Alcohols

Secondary alcohols are oxidized to ketones. There is no further reaction which might complicate things. For example, if you heat the secondary alcohol propan-2-ol with sodium or potassium dichromate(VI) solution acidified with dilute sulphuric acid, you get propanone formed.

Playing around with the reaction conditions makes no difference whatsoever to the product.

Using the simple version of the equation:

$$\underset{\text{propan-2-ol}}{CH_3CH(OH)CH_3} + [O] \longrightarrow \underset{\text{propanone}}{CH_3C(=O)CH_3} + H_2O$$

Using the reaction

The reaction has two uses in testing for aldehydes and ketones.

- First, you can just use it to test for the presence of the carbon-oxygen double bond. You only get an orange or yellow precipitate from a carbon-oxygen double bond in an aldehyde or ketone.
- Secondly, you can use it to help to identify the specific aldehyde or ketone.

The precipitate is filtered and washed with, for example, methanol and then recrystallised from a suitable solvent which will vary depending on the nature of the aldehyde or ketone. For example, you can recrystallise the products from the small aldehydes and ketones from a mixture of ethanol and water.

The crystals are dissolved in the minimum quantity of hot solvent. When the solution cools, the crystals are re-precipitated and can be filtered, washed with a small amount of solvent and dried. They should then be pure.

If you then find the melting point of the crystals, you can compare it with tables of the melting points of 2,4-dinitrophenylhydrazones of all the common aldehydes and ketones to find out which one you are likely to have got.

Ozonolysis of Alkenes

Ozonolysis of Alkenes is used as a synthetic tool for the preparation of aldehydes, ketones, and carboxylic acids depending upon the substitution of the alkene. It is also a structural analytical tool called a de-gradative analysis in which the location of Pi bonding is determined within a molecule. This can be done by identifying the fragments produced upon ozonolysis and then piecing the carbonyl fragments back together positioning the Pi bonds between the carbonyl carbons of the fragments. If only one fragment can be identified then the original alkene must have been symmetrical. If two fragments are identified that could mean an unsymmetrical alkene or a diene with one fragment each from the two pi bonds being the same fragment.

Let's concentrate upon the synthetic value of this reaction. Treatment of an alkene with alkyl groups bonded to at least one of the sp2 carbons of the Pi bond at -60°C will produce the ozonide intermediate. Treatment of the ozonide with Zn and water will produce the fragmentation to produce at least one ketone fragment.

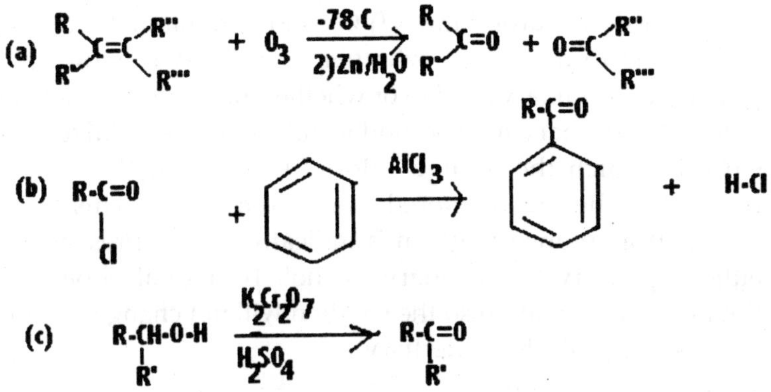

Fig 1: Preparations of ketones

Friedel Crafts Acylation of Benzene

The reaction involves a Lewis acid $AlCl_3$ as a catalyst. An acyl halide is reacted with a benzene ring. The catalyst weakens the bond between the acyl (carbonyl) carbon and the Halogen atom with the bonding electrons kept by the halogen atom. The resulting acylium ion then reacts with the Benzene ring in an electrophilic substitution. (See Fig 1(b)above). This produces an aromatic ketone. This is often the prerequisite to synthesizing an alkyl benzene since the resulting acyl benzene can be reduced to an alkyl benzene using HCl and Carbon Monoxide.

Oxidation of a Secondary Alcohol

Secondary alcohols may be oxidized using any moderate to strong oxidizing agent to a ketone. This is similar to a dehydrogenation where Hydrogen atoms are eliminated at the carbanol carbon and the Hydroxylic Oxygen to produce the Pi bond of the carbonyl center.(See Fig 1-c above) The oxidizing agent is Chromic Acid, H_2CrO_4 which can be prepared as needed by adding Sulfuric Acid to a Potassium Dichromate solution in which the alcohol has been added. Indeed, this reaction is usually carried out in acetone or acetic acid solutions.

The use of Chromium (VI) Oxide (Chromium Tri-oxide) allows one to determine whether an unknown is either a primary or secondary alcohol or whether the suspected alcohol is tertiary. Primary and Secondary alcohols are oxidized by CrO_3 in which the orange colored Chromium (VI) ion is reduced to the green colored Chromium III ion. This color change from orange to green is indicative of the presence of either a primary or secondary alcohol. Tertiary alcohols will not undergo Oxidation so the solution will not change color if the suspected alcohol is tertiary.

Hydration of an Alkyne

This is a synthesis that occurred as a historical "accident". There have been many times in history where important discoveries were made as a result of an accident or mistake that was made in the laboratory. In this case a research team was trying unsuccessfully to have an alkyne undergo hydration using Sulfuric Acid as a catalysts, but the yields were extremely poor with no detectable product being formed. One day an accident occurred in the laboratory when one of the mercury thermometers was accidentally broken. Some of the Mercury entered the reaction mixture and immediately a reaction began!! Upon further investigation, it was concluded that the entrance of the free Mercury formed Mercury II Sulfate by reacting with the Sulfuric Acid. Mercury II Sulfate turned out to be an excellent catalyst for the Hydration reaction!! The products were just as strange as the occurrence of the reaction itself. There were actually two products that occurred which differed by only a movement of a pair of bonding electrons and a proton. These two products were called "tautomers" and this involved the formation of an enol which rearranged into a keto (ketone) form.(See Fig 2-a below) It turns out that the ketone is the more stable of the two tautomers. The equilibrium can be forced to shift toward the ketone side by boiling off the ketone which has a lower boiling point than the enol tautomer. As the ketone is removed that shifts the equilibrium further to the keto side according to LeChatlier's Principle.

$$\text{(a)}\quad CH_3\text{-}C{\equiv}CH + H_2O \xrightarrow[HgSO_4]{H_2SO_4} \underset{\text{enol}}{CH_3\text{-}\underset{OH}{C}{=}CH_2} \rightleftarrows \underset{\text{keto}}{CH_3\text{-}\underset{CH_3}{C}{=}O}$$

$$\text{(b)}\quad R\text{-}\underset{Cl}{C}{=}O + LiCu(R')_2 \longrightarrow R\text{-}\underset{R'}{C}{=}O$$

$$\text{(c)}\quad R'\overset{\cdot\,+}{Mg}Br + R\text{-}C{\equiv}N \longrightarrow R\text{-}\underset{R'}{C}{=}N \xrightarrow[H_3O^+]{H_2O} R\text{-}\underset{R'}{C}{=}O$$

Fig. 2: Preparation of ketones

PHYSICAL PROPERTIES

Boiling Points

None of the Hydrogen atoms connected to an aldehyde or ketone are bonded to an Oxygen or Nitrogen so they do not attract other molecules with the strong Hydrogen bonding. For this reason the aldehydes do not have as high a boiling point for the same sized alcohol that does have a Hydrogen bonded to an Oxygen. However aldehydes and ketones do have the carbonyl structure which is polar since the Oxygen is much higher in electronegativity than carbon atom. Therefore aldehydes and ketones will exhibit dipole-dipole interactions as well as the weak London dispersion intermolecular forces which make them have higher boiling points compared to the hydrocarbons and the ethers. For example, propanal and acetone have boiling points of 49°C and 56°C respectively. The difference is due to the slightly higher molecular mass of acetone. This is to be compared with 1-propanol of 97°C and Ethyl Methyl Ether at 8°C and butane at 0°C. The extra strong hydrogen bonding between the 1-Propanol molecules would account for its higher boiling point. Methanal is a gas (boiling point -21°C), and ethanal has a boiling point of +21°C. That means that ethanal boils at close to room temperature.

The other aldehydes and the ketones are liquids, with boiling points rising as the molecules get bigger.

The size of the boiling point is governed by the strengths of the intermolecular forces.

Van der Waals Dispersion Forces

These attractions get stronger as the molecules get longer and have more electrons. That increases the sizes of the temporary dipoles that are set up. This is why the boiling points increase as the number of carbon atoms in the chains increases - irrespective of whether you are talking about aldehydes or ketones.

Van der Waals dipole-dipole attractions

Both aldehydes and ketones are polar molecules because of the presence of the carbon-oxygen double bond. As well as the dispersion forces, there will also be attractions between the permanent dipoles on nearby molecules.

That means that the boiling points will be higher than those of similarly sized hydrocarbons - which only have dispersion forces.

It is interesting to compare three similarly sized molecules. They have similar lengths, and similar (although not identical) numbers of electrons.

molecule	*type*	*boiling point (°C)*
$CH_3CH_2CH_3$	alkane	-42
CH_3CHO	aldehyde	+21
CH_3CH_2OH	alcohol	+78

Notice that the aldehyde (with dipole-dipole attractions as well as dispersion forces) has a boiling point higher than the similarly sized alkane which only has dispersion forces.

However, the aldehyde's boiling point isn't as high as the alcohol's. In the alcohol, there is hydrogen bonding as well as the other two kinds of intermolecular attraction.

Although the aldehydes and ketones are highly polar molecules, they don't have any hydrogen atoms attached directly to the oxygen, and so they can't hydrogen bond with each other.

Water Solubility of Aldehydes and Ketones

The carbonyl Oxygen with its lone pairs allow for ketones and aldehydes to hydrogen bond with Hydroxylic compounds like water and alcohols. Therefore the low molecular mass (up to four carbons) of aldehydes and ketones allow them to be very soluble in water. This solubility is similar to alcohols and ethers which also have Oxygen atoms with lone pairs of electrons. Aldehydes and ketones of greater than five carbons generally will not be soluble in water as are the alcohols. This is because the increased size of the hydrocarbon portion will prevent water molecules from being attracted to the organic molecules. In other words the solvation process is hampered and the water molecules are not capable of surrounding each organic molecule and separating them. The polar water molecules have little attraction for hydrocarbons.

Solubility in Water

The small aldehydes and ketones are freely soluble in water but solubility falls with chain length. For example, methanal, ethanal and propanone - the common small aldehydes and ketones - are miscible with water in all proportions.

The reason for the solubility is that although aldehydes and ketones can't hydrogen bond with themselves, they *can* hydrogen bond with water molecules.

One of the slightly positive hydrogen atoms in a water molecule can be sufficiently attracted to one of the lone pairs on the oxygen atom of an aldehyde or ketone for a hydrogen bond to be formed.

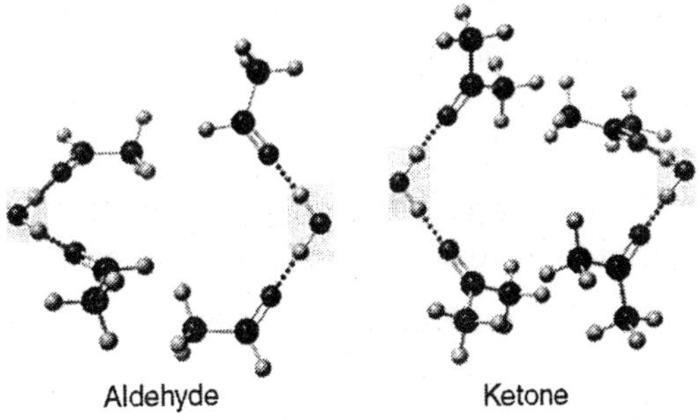

Fig. Hydrogen bonding in aldehydes & ketones

There will also, of course, be dispersion forces and dipole-dipole attractions between the aldehyde or ketone and the water molecules.

Forming these attractions releases energy which helps to supply the energy needed to separate the water molecules and aldehyde or ketone molecules from each other before they can mix together.

As chain lengths increase, the hydrocarbon "tails" of the molecules (all the hydrocarbon bits apart from the carbonyl group) start to get in the way.

By forcing themselves between water molecules, they break the relatively strong hydrogen bonds between water molecules without replacing them by anything as good. This makes the process energetically less profitable, and so solubility decreases.

ADDITION-ELIMINATION REACTIONS OF ALDEHYDES AND KETONES

Important Reactions of the Carbonyl Group

The slightly positive carbon atom in the carbonyl group can be attacked by *nucleophiles*. A nucleophile is a negatively charged ion (for example, a cyanide ion, CN^-), or a slightly negatively charged part of a molecule (for example, the lone pair on a nitrogen atom in ammonia, NH_3).

During the reaction, the carbon-oxygen double bond gets broken. The net effect of all this is that the carbonyl group undergoes *addition reactions*, often followed by the loss of a water molecule. This gives a reaction known as *addition-elimination* or *condensation*.

Both aldehydes and ketones contain a carbonyl group. That means that their reactions are very similar in this respect.

Where Aldehydes and Ketones Differ

An aldehyde differs from a ketone by having a hydrogen atom attached to the carbonyl group. This makes the aldehydes very easy to oxidize.

For example, ethanal, CH_3CHO, is very easily oxidized to either ethanoic acid, CH_3COOH, or ethanoate ions, CH_3COO^-.

Ketones don't have that hydrogen atom and are resistant to oxidation. They are only oxidized by powerful oxidising agents which have the ability to break carbon-carbon bonds.

SIMPLE ADDITION TO ALDEHYDES AND KETONES

Addition of Hydrogen Cyanide to Aldehydes and Ketones

Hydrogen cyanide adds across the carbon-oxygen double bond in aldehydes and ketones to produce compounds known as hydroxynitriles. These used to be known as cyanohydrins.

For example, with ethanal (an aldehyde) you get 2-hydroxypropanenitrile:

$$CH_3-C(=O)-H + HCN \longrightarrow CH_3-C(OH)(H)-CN$$

With propanone (a ketone) you get 2-hydroxy-2-methylpropanenitrile:

$$(CH_3)_2C=O + HCN \longrightarrow CH_3-C(OH)(CH_3)-CN$$

The reaction isn't normally done using hydrogen cyanide itself, because this is an extremely poisonous gas. Instead, the aldehyde or ketone is mixed with a solution of sodium or potassium cyanide in water to which a little sulphuric acid has been added. The pH of the solution is adjusted to about 4 - 5, because this gives the fastest reaction. The reaction happens at room temperature.

The solution will contain hydrogen cyanide (from the reaction between the sodium or potassium cyanide and the sulphuric acid), but still contains some free cyanide ions. This is important for the mechanism.

Uses of the Reaction

The product molecules contain two functional groups:

- The -OH group which behaves like a simple alcohol and can be replaced by other things like chlorine, which can in turn be replaced to give, for example, an -NH2 group;
- The -CN group which is easily converted into a carboxylic acid group -COOH.

For example, starting from a hydroxynitrile made from an aldehyde, you can quite easily produce relatively complicated molecules like 2-amino acids - the amino acids which are used to construct proteins.

$$R\text{-}CH(OH)\text{-}CN \longrightarrow R\text{-}CH(Cl)\text{-}CN \longrightarrow R\text{-}CH(NH_2)\text{-}CN \longrightarrow R\text{-}CH(NH_2)\text{-}COOH$$

Addition of sodium hydrogensulphite to aldehydes and ketones

Sodium hydrogensulphite used to be known as sodium bisulphite, and you might well still come across it in organic textbooks under this name - or using the bisulfite spelling.

This reaction only works well for aldehydes. In the case of ketones, one of the hydrocarbon groups attached to the carbonyl group needs to be a methyl group. Bulky groups attached to the carbonyl group get in the way of the reaction happening.

The aldehyde or ketone is shaken with a saturated solution of sodium hydrogensulphite in water. Where the product is formed, it separates as white crystals.

In the case of ethanal, the equation is:

$$CH_3\text{-}CHO + Na^+\ HSO_3^- \longrightarrow CH_3\text{-}CH(OH)\text{-}SO_3^-\ Na^+$$

and with propanone, the equation is:

$$CH_3-C(=O)-CH_3 + Na^+ HSO_3^- \longrightarrow CH_3-C(OH)(CH_3)-SO_3^- Na^+$$

These compounds are rarely named systematically, and are usually known as "hydrogensulphite (or bisulphite) addition compounds".

Uses of the Reaction

The reaction is usually used during the purification of aldehydes (and any ketones that it works for). The addition compound can be split easily to regenerate the aldehyde or ketone by treating it with either dilute acid or dilute alkali.

If you have an impure aldehyde, for example, you could shake it with a saturated solution of sodium hydrogensulphite to produce the crystals. These crystals could easily be filtered and washed to remove any other impurities. Addition of dilute acid, for example, would then regenerate the original aldehyde.

The Reaction with 2,4-dinitrophenylhydrazine

2,4-dinitrophenylhydrazine is often abbreviated to 2,4-DNP or 2,4-DNPH. A solution of 2,4-dinitrophenylhydrazine in a mixture of methanol and sulphuric acid is known as Brady's reagent.

Hydrazine is:

$$H_2N\text{-}NH_2$$

hydrazine

In phenylhydrazine, one of the hydrogens is replaced by a phenyl group, C_6H_5. This is based on a benzene ring.

$$H_2N-NH-C_6H_5$$

phenylhydrazine

In 2,4-dinitrophenylhydrazine, there are two nitro groups, NO_2, attached to the phenyl group in the 2- and 4- positions.

The corner with the nitrogen attached is counted as the number 1 position, and you just number clockwise around the ring.

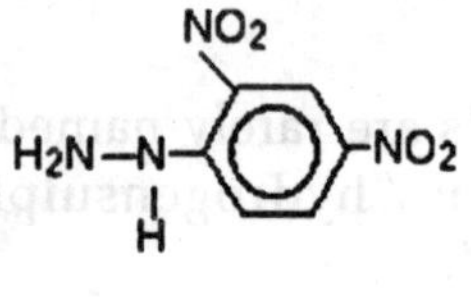

2,4-dinitrophenylhydrazine

Doing the Reaction

Details vary slightly depending on the nature of the aldehyde or ketone, and the solvent that the 2,4-dinitrophenylhydrazine is dissolved in. Assuming you are using Brady's reagent (a solution of the 2,4-dinitrophenylhydrazine in methanol and sulphuric acid):

Add either a few drops of the aldehyde or ketone, or possibly a solution of the aldehyde or ketone in methanol, to the Brady's reagent. A bright orange or yellow precipitate shows the presence of the carbon-oxygen double bond in an aldehyde or ketone.

This is the simplest test for an aldehyde or ketone.

The Chemistry of the Reaction

The overall reaction is given by the equation:

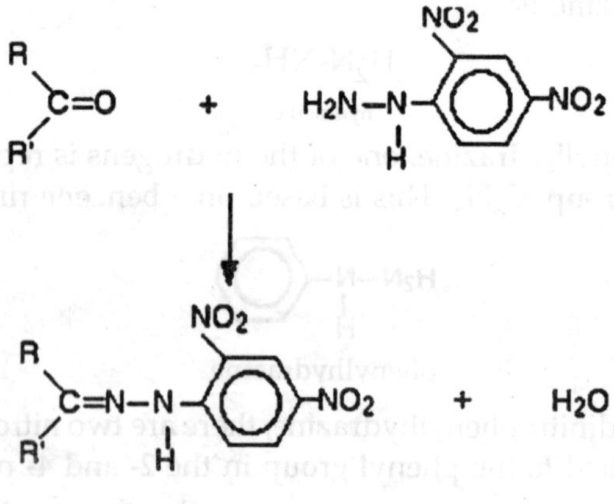

R and R′ can be any combination of hydrogen or hydrocarbon groups (such as alkyl groups). If at least one of them is a hydrogen, then the original compound is an aldehyde. If both are hydrocarbon groups, then it is a ketone.

Look carefully at what has happened.

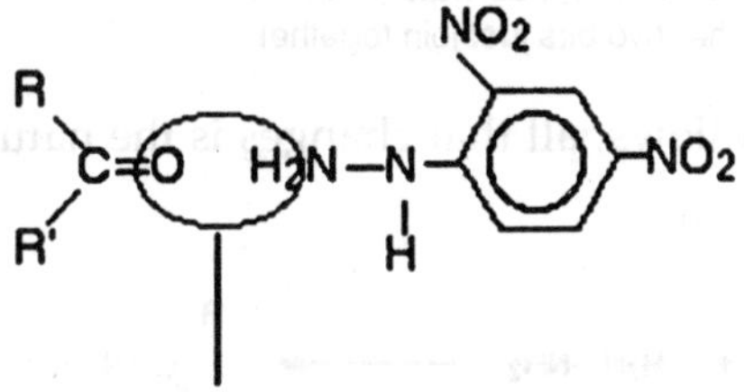

$$\begin{matrix} R \\ & C{=}O & + & H_2N{-}NH{-}C_6H_3(NO_2)_2 \\ R' \end{matrix}$$

This gets lost as water, and the other two bits just join together

The product is known as a "2,4-dinitrophenylhydrazone". Notice that all that has changed is the ending from "-ine" to "-one". That's possibly confusing.

The product from the reaction with ethanal would be called ethanal 2,4-dinitrophenylhydrazone; from propanone, you would get propanone 2,4-dinitrophenylhydrazone - and so on. That's not too difficult!

The reaction is known as a *condensation reaction*. A condensation reaction is one in which two molecules join together with the loss of a small molecule in the process. In this case, that small molecule is water.

In terms of mechanisms, this is a *nucleophilic addition-elimination* reaction. The 2,4-dinitrophenylhydrazine first adds across the carbon-oxygen double bond (the addition stage) to give an intermediate compound which then loses a molecule of water (the elimination stage).

Some other Similar Reactions

If you go back and look at the equations, nothing in the 2,4-dinitrophenylhydrazine changes during the reaction apart from the $-NH_2$ group. You can get a similar reaction if the $-NH_2$ group is attached to other things.

In each case, the reaction would look like this:

$$R(R')C{=}O + H_2N{-}X \longrightarrow R(R')C{=}N{-}X + H_2O$$

This gets lost as water, and the other two bits just join together

In what follows, all that changes is the nature of the "X".

with hydrazine

$$R(R')C{=}O + H_2N{-}NH_2 \longrightarrow R(R')C{=}N{-}NH_2 + H_2O$$

The product is a "hydrazone". If you started from propanone, it would be propanone hydrazone.

with phenylhydrazine

$$R(R')C{=}O + H_2N{-}NH{-}C_6H_5 \longrightarrow R(R')C{=}N{-}NH{-}C_6H_5 + H_2O$$

The product is a "phenylhydrazone".

with hydroxylamine

$$R(R')C{=}O + H_2N{-}OH \longrightarrow R(R')C{=}N{-}OH + H_2O$$

The product is an "oxime" - for example, ethanal oxime.

REDUCTION

Here we are going to discuss the reduction of aldehydes and ketones by two similar reducing agents - lithium tetrahydridoaluminate(III) (also known as lithium aluminium hydride) and sodium tetrahydridoborate(III) (sodium borohydride).

Reducing Agents

Despite the fearsome names, the structures of the two reducing agents are very simple. In each case, there are four hydrogens ("tetrahydido") around either aluminium or boron in a negative ion (shown by the "ate" ending).

The "(III)" shows the oxidation state of the aluminium or boron, and is often left out because these elements only ever show the +3 oxidation state in their compounds. The formulae of the two compounds are $LiAlH_4$ and $NaBH_4$.

Their structures are:

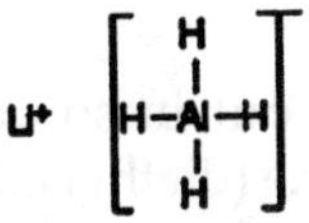

$$Li^+ \left[\begin{array}{c} H \\ | \\ H-Al-H \\ | \\ H \end{array} \right]^- \qquad Na^+ \left[\begin{array}{c} H \\ | \\ H-B-H \\ | \\ H \end{array} \right]^-$$

lithium tetrahydridoaluminate 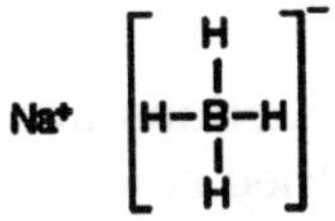 — sodium tetrahydridoborate

In each of the negative ions, one of the bonds is a co-ordinate covalent (dative covalent) bond using the lone pair on a hydride ion (H^-) to form a bond with an empty orbital on the aluminium or boron.

Reduction of an Aldehyde

You get exactly the same organic product whether you use lithium tetrahydridoaluminate or sodium tetrahydridoborate.

For example, with ethanal you get ethanol:

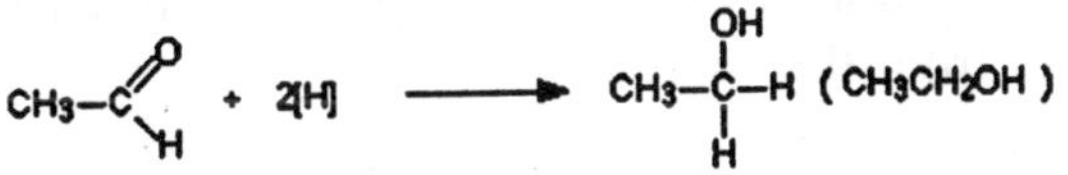

$$CH_3-CHO + 2[H] \longrightarrow CH_3-\underset{H}{\overset{OH}{C}}-H \; (CH_3CH_2OH)$$

Notice that this is a simplified equation. [H] means "hydrogen from a reducing agent".

In general terms, reduction of an aldehyde leads to a *primary alcohol.*

Reduction of a Ketone

Again the product is the same whichever of the two reducing agents you use.

For example, with propanone you get propan-2-ol:

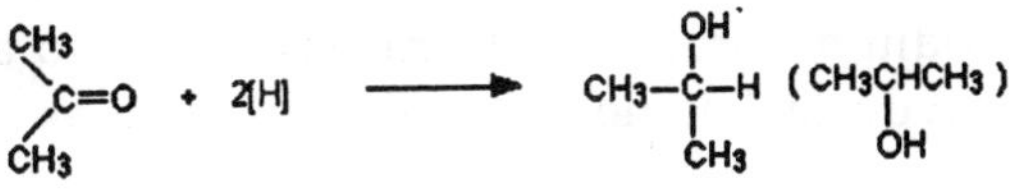

$$(CH_3)_2C=O + 2[H] \longrightarrow CH_3-\underset{CH_3}{\overset{OH}{C}}-H \; (CH_3\underset{OH}{C}HCH_3)$$

Reduction of a ketone leads to a *secondary alcohol.*

Reaction Details

Using lithium tetrahydridoaluminate (lithium aluminium hydride)

Lithium tetrahydridoaluminate is much more reactive than sodium tetrahydridoborate. It reacts violently with water and alcohols, and so any reaction must exclude these common solvents.

The reactions are usually carried out in solution in a carefully dried ether such as ethoxyethane (diethyl ether). The reaction happens at room temperature, and takes place in two separate stages.

In the first stage, a salt is formed containing a complex aluminium ion. The following equations show what happens if you start with a general aldehyde or ketone. R and R′ can be any combination of hydrogen or alkyl groups.

$$4\,R\text{-}C(=O)\text{-}R' + Li^+\,AlH_4^- \longrightarrow \left(R'\text{-}\underset{H}{\overset{R}{C}}\text{-}O\right)_4 Al^-\,Li^+$$

The product is then treated with a dilute acid (such as dilute sulphuric acid or dilute hydrochloric acid) to release the alcohol from the complex ion.

$$\left(R'\text{-}\underset{H}{\overset{R}{C}}\text{-}O\right)_4 Al^- + 4H^+_{(aq)} \longrightarrow 4\,R'\text{-}\underset{H}{\overset{R}{C}}\text{-}OH + Al^{3+}_{(aq)}$$

The alcohol formed can be recovered from the mixture by fractional distillation.

Using Sodium Tetrahydridoborate (sodium borohydride)

Sodium tetrahydridoborate is a more gentle (and therefore safer) reagent than lithium tetrahydridoaluminate. It can be used in solution in alcohols or even solution in water - provided the solution is alkaline.

Solid sodium tetrahydridoborate is added to a solution of the aldehyde or ketone in an alcohol such as methanol,

ethanol or propan-2-ol. Depending on which recipe you read, it is either heated under reflux or left for some time around room temperature. This almost certainly varies depending on the nature of the aldehyde or ketone.

At the end of this time, a complex similar to the previous one is formed.

$$4\ R\text{-}C(=O)\text{-}R' + Na^+\ BH_4^- \longrightarrow \left(R'\text{-}\overset{R}{\underset{H}{C}}\text{-}O\text{-}\right)_4 B^-\ Na^+$$

In the second stage of the reaction, water is added and the mixture is boiled to release the alcohol from the complex.

$$\left(R'\text{-}\overset{R}{\underset{H}{C}}\text{-}O\text{-}\right)_4 B^- + 3H_2O \longrightarrow 4\ R'\text{-}\overset{R}{\underset{H}{C}}\text{-}O\text{-}H + NaH_2BO_3$$

Again, the alcohol formed can be recovered from the mixture by fractional distillation.

REACTION OF ALDEHYDES AND KETONES WITH GRIGNARD REAGENTS

These are reactions of the carbon-oxygen double bond, and so aldehydes and ketones react in exactly the same way - all that changes are the groups that happen to be attached to the carbon-oxygen double bond.

It is much easier to understand what is going on by looking closely at the general case (using "R" groups rather than specific groups) - and then slotting in the various real groups as and when you need to. The "R" groups can be either hydrogen or alkyl in any combination.

In the first stage, the Grignard reagent adds across the carbon-oxygen double bond:

$$CH_3CH_2MgBr + R\text{—}C(=O)\text{—}R' \longrightarrow CH_3CH_2\text{-}\overset{R}{\underset{R'}{C}}\text{-}O\text{-}MgBr$$

Dilute acid is then added to this to hydrolyze it.

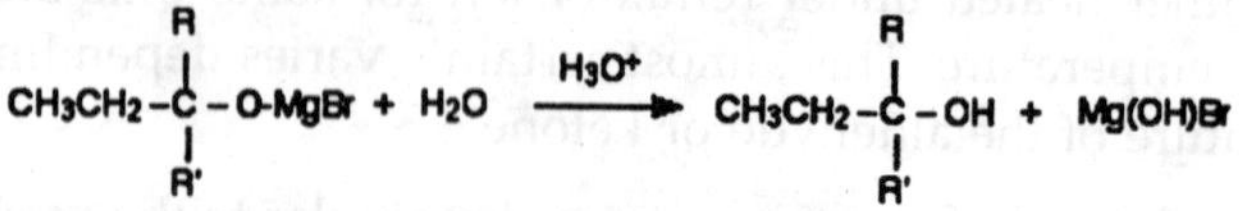

$$CH_3CH_2-\underset{R'}{\overset{R}{C}}-O\text{-}MgBr + H_2O \xrightarrow{H_3O^+} CH_3CH_2-\underset{R'}{\overset{R}{C}}-OH + Mg(OH)Br$$

An alcohol is formed. One of the key uses of Grignard reagents is the ability to make complicated alcohols easily.

What sort of alcohol you get depends on the carbonyl compound you started with - in other words, what R and R' are.

The Reaction between Grignard reagents and methanal

In methanal, both R groups are hydrogen. Methanal is the simplest possible aldehyde.

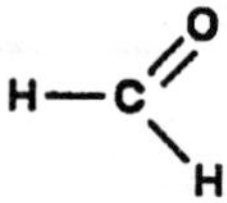

$$H-C(=O)-H$$

methanal

Assuming that you are starting with CH_3CH_2MgBr and using the general equation above, the alcohol you get always has the form:

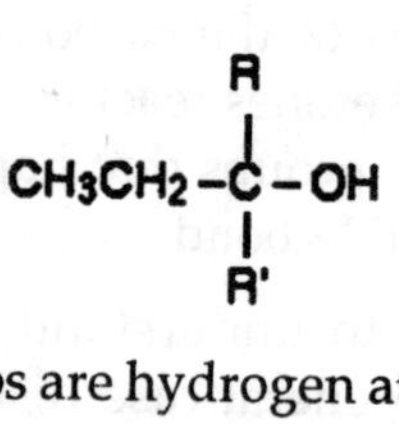

$$CH_3CH_2-\underset{R'}{\overset{R}{C}}-OH$$

Since both R groups are hydrogen atoms, the final product will be:

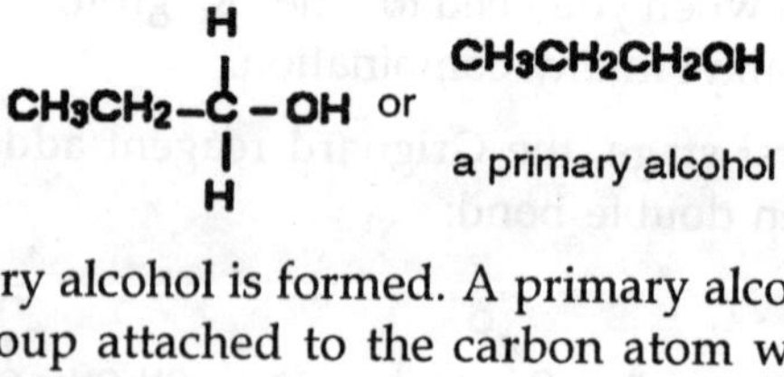

$$CH_3CH_2-\underset{H}{\overset{H}{C}}-OH \quad \text{or} \quad CH_3CH_2CH_2OH$$

a primary alcohol

A primary alcohol is formed. A primary alcohol has only one alkyl group attached to the carbon atom with the -OH group on it.

You could obviously get a different primary alcohol if you started from a different Grignard reagent.

The Reaction between Grignard reagents and other aldehydes

The next biggest aldehyde is ethanal. One of the R groups is hydrogen and the other CH_3.

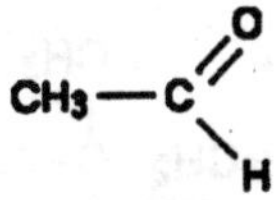

$$CH_3-C(=O)-H$$

ethanal

Again, think about how that relates to the general case. The alcohol formed is:

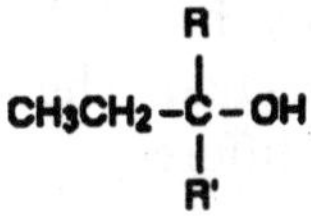

$$CH_3CH_2-C(R)(R')-OH$$

So this time the final product has one CH_3 group and one hydrogen attached:

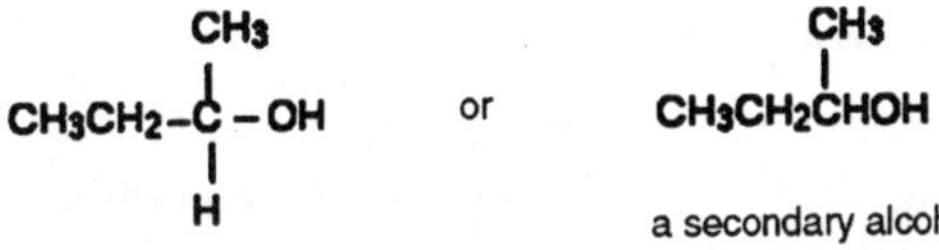

$$CH_3CH_2-C(CH_3)(H)-OH \quad \text{or} \quad CH_3CH_2CH(CH_3)OH$$

a secondary alcoho

A secondary alcohol has two alkyl groups (the same or different) attached to the carbon with the -OH group on it.

You could change the nature of the final secondary alcohol by either:

- Changing the nature of the Grignard reagent - which would change the CH_3CH_2 group into some other alkyl group;
- Changing the nature of the aldehyde - which would change the CH_3 group into some other alkyl group.

The reaction between Grignard reagents and ketones

Ketones have two alkyl groups attached to the carbon-oxygen double bond. The simplest one is propanone.

$$CH_3-\overset{\displaystyle O}{\overset{\|}{C}}-CH_3$$

propanone

This time when you replace the R groups in the general formula for the alcohol produced you get a tertiary alcohol.

$$CH_3CH_2-\underset{\displaystyle CH_3}{\underset{|}{\overset{\displaystyle CH_3}{\overset{|}{C}}}}-OH$$

a tertiary alcohol

A tertiary alcohol has three alkyl groups attached to the carbon with the -OH attached. The alkyl groups can be any combination of same or different.

You could ring the changes on the product by

- Changing the nature of the Grignard reagent - which would change the CH_3CH_2 group into some other alkyl group;
- Changing the nature of the ketone - which would change the CH_3 groups into whatever other alkyl groups you choose to have in the original ketone.

OXIDATION OF ALDEHYDES AND KETONES

Here we will look at ways of distinguishing between aldehydes and ketones using oxidising agents such as acidified potassium dichromate(VI) solution, Tollens' reagent, Fehling's solution and Benedict's solution.

Why do Aldehydes and Ketones behave Differently?

You will remember that the difference between an aldehyde and a ketone is the presence of a hydrogen atom attached to the carbon-oxygen double bond in the aldehyde. Ketones don't have that hydrogen.

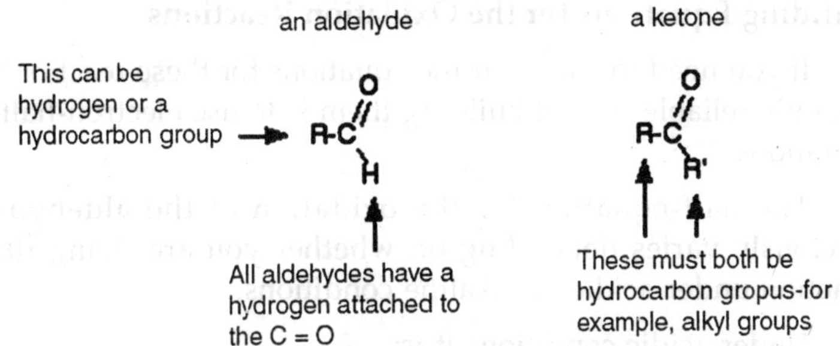

The presence of that hydrogen atom makes aldehydes very easy to oxidize. Or, put another way, they are strong reducing agents.

Because ketones don't have that particular hydrogen atom, they are resistant to oxidation. Only very strong oxidising agents like potassium manganate(VII) solution (potassium permanganate solution) oxidize ketones - and they do it in a destructive way, breaking carbon-carbon bonds.

Provided you avoid using these powerful oxidising agents, you can easily tell the difference between an aldehyde and a ketone. Aldehydes are easily oxidised by all sorts of different oxidising agents: ketones aren't.

What is Formed when Aldehydes are Oxidised?

It depends on whether the reaction is done under acidic or alkaline conditions. Under acidic conditions, the aldehyde is oxidised to a carboxylic acid. Under alkaline conditions, this couldn't form because it would react with the alkali. A salt is formed instead.

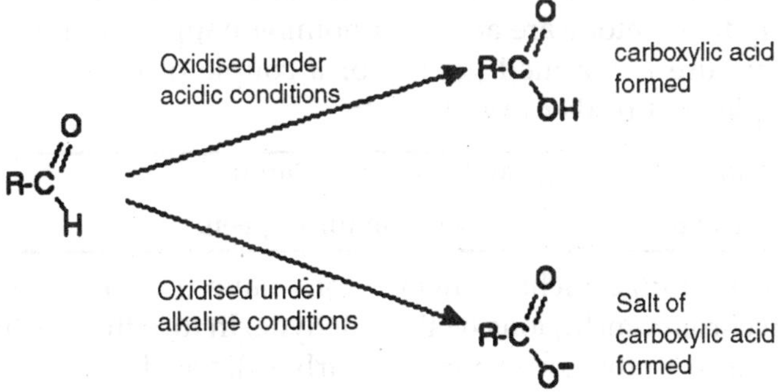

Building Equations for the Oxidation Reactions

If you need to work out the equations for these reactions, the only reliable way of building them is to use electron-half-equations.

The half-equation for the oxidation of the aldehyde obviously varies depending on whether you are doing the reaction under acidic or alkaline conditions.

Under acidic conditions it is:

$$RCHO + H_2O \longrightarrow RCOOH + 2H^+ + 2e^-$$

and under alkaline conditions:

$$RCHO + 3OH^- \longrightarrow RCOO^- + 2H_2O + 2e^-$$

These half-equations are then combined with the half-equations from whatever oxidising agent you are using. Examples are given in detail below.

Specific Examples

In each of the following examples, we are assuming that you know that you have either an aldehyde or a ketone. There are lots of other things which could also give positive results.

Assuming that you know it has to be one or the other, in each case, a ketone does nothing. Only an aldehyde gives a positive result.

Using Acidified Potassium Dichromate(VI) solution

A small amount of potassium dichromate(VI) solution is acidified with dilute sulphuric acid and a few drops of the aldehyde or ketone are added. If nothing happens in the cold, the mixture is warmed gently for a couple of minutes - for example, in a beaker of hot water.

ketone	No change in the orange solution.
aldehyde	Orange solution turns green.

The orange dichromate(VI) ions have been reduced to green chromium(III) ions by the aldehyde. In turn the aldehyde is oxidised to the corresponding carboxylic acid.

The electron-half-equation for the reduction of dichromate(VI) ions is:

Cr_2O7^{2-} + $14H^+$ +$6e^-$ ⟶ $2Cr^{3+}$ + $7H_2O$

Combining that with the half-equation for the oxidation of an aldehyde under acidic conditions:

RCHO + H_2O ⟶ RCOOH + $2H^+$ + $2e^-$

gives the overall equation:

3RCHO + $Cr_2O_7^{2-}$ + BH^+ ⟶ 3RCOOH + $2Cr^{3+}$ + $4H_2O$

Using Tollens' reagent (the silver mirror test)

Tollens' reagent contains the diamminesilver(I) ion, $[Ag(NH_3)_2]^+$.

This is made from silver(I) nitrate solution. You add a drop of sodium hydroxide solution to give a precipitate of silver(I) oxide, and then add just enough dilute ammonia solution to redissolve the precipitate.

To carry out the test, you add a few drops of the aldehyde or ketone to the freshly prepared reagent, and warm gently in a hot water bath for a few minutes.

Ketone	No change in the colourless solution.
Aldehyde	The colourless solution produces a grey precipitate of silver, or a silver mirror on the test tube.

Aldehydes reduce the diamminesilver(I) ion to metallic silver. Because the solution is alkaline, the aldehyde itself is oxidised to a salt of the corresponding carboxylic acid.

The electron-half-equation for the reduction of the diamminesilver(I) ions to silver is:

$Ag(NH_3)^{2+}$ + e^- ⟶ Ag + $2NH_3$

Combining that with the half-equation for the oxidation of an aldehyde under alkaline conditions:

RCHO + $3OH^-$ ⟶ $RCOO^-$ + $2H_2O$ + $2e^-$

gives the overall equation:

$$2Ag(NH_3)_2^+ + RCHO + 3OH^- \longrightarrow 2Ag + RCOO^- + 4NH_3 + 2H_2O$$

Using Fehling's Solution or Benedict's solution

Fehling's solution and Benedict's solution are variants of essentially the same thing. Both contain complexed copper(II) ions in an alkaline solution.

Fehling's solution contains copper(II) ions complexed with tartrate ions in sodium hydroxide solution. Complexing the copper(II) ions with tartrate ions prevents precipitation of copper(II) hydroxide.

Benedict's solution contains copper(II) ions complexed with citrate ions in sodium carbonate solution. Again, complexing the copper(II) ions prevents the formation of a precipitate - this time of copper(II) carbonate.

Both solutions are used in the same way. A few drops of the aldehyde or ketone are added to the reagent, and the mixture is warmed gently in a hot water bath for a few minutes.

ketone	No change in the blue solution.
aldehyde	The blue solution produces a dark red precipitate of copper(I) oxide.

Aldehydes reduce the complexed copper(II) ion to copper(I) oxide. Because the solution is alkaline, the aldehyde itself is oxidised to a salt of the corresponding carboxylic acid.

The equations for these reactions are always simplified to avoid having to write in the formulae for the tartrate or citrate ions in the copper complexes. The electron-half-equations for both Fehling's solution and Benedict's solution can be written as:

$$2Cu^{2+}(\text{in complex}) + 2OH^- + 2e^- \longrightarrow Cu_2O + H_2O$$

Combining that with the half-equation for the oxidation of an aldehyde under alkaline conditions:

$RCHO + 3OH^- \longrightarrow RCOO^- + 2H_2O + 2e^-$

gives the overall equation:

$RCHO + 2Cu^{2+}$(in complex) $+ 5OH^- \longrightarrow RCOO^- + Cu_2O + 3H_2O$

THE TRIIODOMETHANE (IODOFORM) REACTION WITH ALDEHYDES AND KETONES

The triiodomethane (iodoform) reaction can be used to identify the presence of a CH_3CO group in aldehydes and ketones.

Doing the Triiodomethane (iodoform) Reaction

There are two apparently quite different mixtures of reagents that can be used to do this reaction. They are, in fact, chemically equivalent.

Using Iodine and Sodium Hydroxide Solution

This is chemically the more obvious method.

Iodine solution is added to a small amount of aldehyde or ketone, followed by just enough sodium hydroxide solution to remove the colour of the iodine. If nothing happens in the cold, it may be necessary to warm the mixture very gently.

A positive result is the appearance of a very pale yellow precipitate of triiodomethane (previously known as iodoform) - CHI_3.

Apart from its colour, this can be recognized by its faintly "medical" smell. It is used as an antiseptic on the sort of sticky plasters you put on minor cuts, for example.

Using Potassium Iodide and Sodium Chlorate(I) solutions

Sodium chlorate(I) is also known as sodium hypochlorite.

Potassium iodide solution is added to a small amount of aldehyde or ketone, followed by sodium chlorate(I) solution. Again, if no precipitate is formed in the cold, it may be necessary to warm the mixture very gently.

The positive result is the same pale yellow precipitate as before.

The Chemistry of the Triiodomethane (iodoform) reaction

What the triiodomethane (iodoform) reaction shows

A positive result - the pale yellow precipitate of triiodomethane (iodoform) - is given by an aldehyde or ketone containing the grouping:

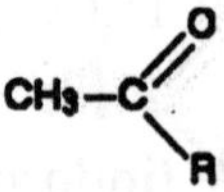

$$CH_3-C(=O)-R$$

"R" can be a hydrogen atom or a hydrocarbon group (for example, an alkyl group).

If "R" is hydrogen, then you have the aldehyde ethanal, CH_3CHO.

- Ethanal is the *only* aldehyde to give the triiodomethane (iodoform) reaction.
- If "R" is a hydrocarbon group, then you have a ketone. Lots of ketones give this reaction, but those that do all have a methyl group on one side of the carbon-oxygen double bond. These are known as *methyl ketones.*

Equations for the Triiodomethane (iodoform) Reaction

We will take the reagents as being iodine and sodium hydroxide solution.

The first stage involves substitution of all three hydrogens in the methyl group by iodine atoms. The presence of hydroxide ions is important for the reaction to happen - they take part in the mechanism for the reaction.

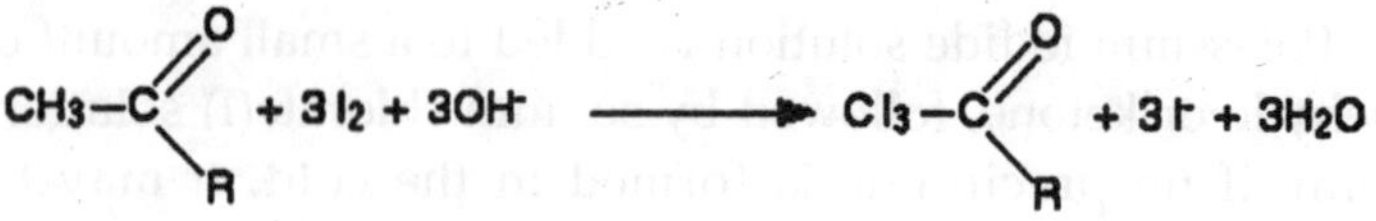

$$CH_3-C(=O)-R + 3I_2 + 3OH^- \longrightarrow CI_3-C(=O)-R + 3I^- + 3H_2O$$

In the second stage, the bond between the CI_3 and the rest of the molecule is broken to produce triiodomethane (iodoform) and the salt of an acid.

$$CI_3-C(=O)R + OH^- \longrightarrow CHI_3 + R-C(=O)O^-$$

This bond is broken

Putting all this together gives the overall equation for the reaction:

$$CH_3-C(=O)R + 3I_2 + 4OH^- \longrightarrow CHI_3 + RCOO^- + 3I^- + 3H_2O$$

Chapter 12

Carboxylic Acids

INTRODUCTION

Carboxylic Acids Contain a -COOH group

The Carboxylic Acid Family is a family of organic compounds with the functional group being the carboxyl group, -COOH. This group is attached to one of the carbons in the rest of the molecule. It is actually a carbonyl group, C=O, bonded to a hydroxyl group, OH. Taking the first four letters of the word carbonyl and the last four letters of the word hydroxyl you get the word carboxyl. This family is a weak acid family reluctantly giving up its proton on the carboxyl group.

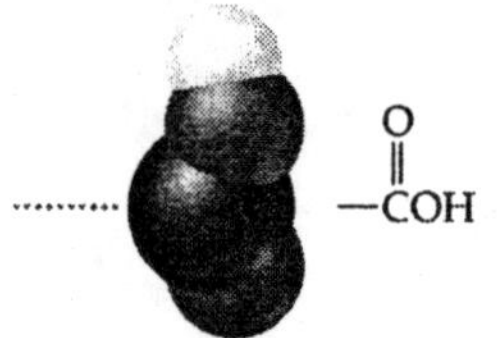

Fig. Carboxylic acid group

Examples of Carboxylic Acids

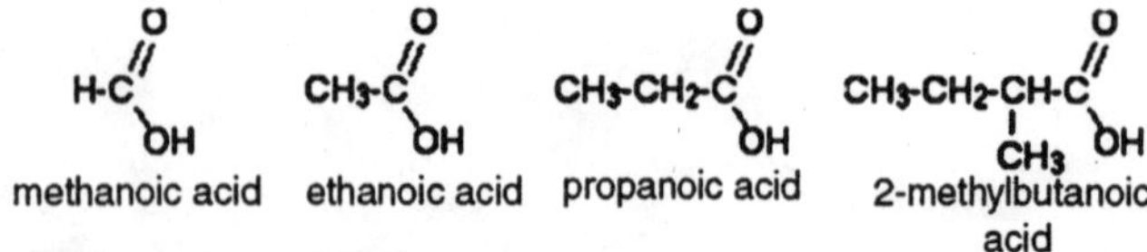

The name counts the total number of carbon atoms in the longest chain - including the one in the -COOH group. If you have side groups attached to the chain, notice that you always count from the carbon atom in the -COOH group as being number 1.

IUPAC RULES FOR NAMING CARBOXYLIC ACIDS

1. Determine the longest continuous chain of carbons that have the Carbonyl carbon as part of the continuous chain. By "longest continuous chain" is meant to be able to trace through the carbons without raising the tracer (or finger) off the surface. The chain does not necessarily have to be straight.
2. Number the carbons in the chain beginning with the carbonyl carbon as the C-1 carbon. This rule is similar to the rule found in the naming of aldehydes.
3. Identify the various branching groups attached to this continuous chain of carbons by name
4. Name the branched groups in alphabetical order attaching (hyphenating) the carbon number it is attached to along the continuous chain of carbons to the front of the branch name. If more than one of the same kind of branched group is attached to the chain, identify the number carbon each group is attached to as a series of numbers separated by commas between each number then a hyphen and finally use a greek prefix attached to the branch name.
5. If it is a saturated acid use the normal alkane name corresponding to the number of carbons in the continuous chain. If it has an unsaturation center that is a double bond then use the alkene name and prefix the name with the lowest carbon involving the double bond. If a triple bond is involved then use the alkyne name and prefix the name with the lowest carbon number involving the triple bond
6. Drop the "ane", "ene", or "yne" ending and add the characteristic IUPAC ending for this family which is "oic acid".

Here is an example:

Identify the IUPAC name for the following:

CH_3-CH_2(Cl)-CH_2-CH_2-CH_2-COOH

1. Identify the longest continuous chain of carbons with the carbonyl carbon involved

   Here we have six carbons

2. Number the carbon chain with the carbonyl carbon as carbon #1

   Numbering from right to left.

3. Identify and locate all branched groups ,alphabetically, attaching a prefixed number equal to the carbon number the branch is attached to.

Here we have a chloro group attached to the #5 carbon

So we would have 5-Chloro-

Use the normal alkane name corresponding to six carbons

5-ChloroHexane

change the "ane" ending to "oic acid"

5-ChloroHexanoic Acid

PREPARATION

Making carboxylic acids by oxidising primary alcohols or aldehydes

Chemistry of the Reactions

Primary alcohols and aldehydes are normally oxidized to carboxylic acids using potassium dichromate(VI) solution in the presence of dilute sulphuric acid. During the reaction, the potassium dichromate(VI) solution turns from orange to green.

The potassium dichromate(VI) can just as well be replaced with sodium dichromate(VI). Because what matters is the dichromate(VI) ion, all the equations and colour changes would be identical.

Primary alcohols are oxidized to carboxylic acids in two stages - first to an aldehyde and then to the acid. We often use simplified versions of these equations using "[O]" to represent oxygen from the oxidising agent.

The formation of the aldehyde is shown by the simplified equation:

$$RCH_2OH + [O] \longrightarrow R\text{-}C(=O)H + H_2O$$

This means "oxygen from an oxidising agent"

"R" is a hydrogen atom or a hydrocarbon group such as an alkyl group.

The aldehyde is then oxidized further to give the carboxylic acid:

$$R\text{-}C(=O)H + [O] \longrightarrow R\text{-}C(=O)OH$$

If you start with an aldehyde, you are obviously just doing this second stage.

Starting from the primary alcohol, you could combine these into one single equation to give:

$$RCH_2OH + 2(O) \longrightarrow RCOOH + H_2O$$

For example, if you were converting ethanol into ethanoic acid, the simplified equation would be:

$$CH_3CH_2OH + 2(O) \longrightarrow CH_3COOH + H_2O$$

It is possible that you might want to write proper equations for these reactions rather than these simplified ones. The complete equation for the conversion of a primary alcohol to a carboxylic acid is:

$$3RCH_2OH + 2Cr_2O_7^{2-} + 16H^+ \longrightarrow 3RCOOH + 4Cr^{3+} + 11H_2O$$

or if you were starting from an aldehyde is:

$$3RCHO + Cr_2O_7^{2-} + BH^+ \longrightarrow 3RCOOH + 2Cr^{3+} + 4H_2O$$

It would actually be quite uncommon to make an acid starting from an aldehyde, but very common to start from a primary alcohol. The conversion of ethanol into ethanoic acid would be a typical example.

The alcohol is heated under reflux with an excess of a mixture of potassium dichromate(VI) solution and dilute sulphuric acid.

Heating under reflux (heating in a flask with a condenser placed vertically in it) prevents any aldehyde formed escaping before it has time to be oxidized to the carboxylic acid.

Using an excess of oxidising agent is to be sure that there is enough oxidising agent present for the oxidation to go all the way to the carboxylic acid.

When oxidation is complete, the mixture can be distilled. You end up with an aqueous solution of the acid.

MAKING CARBOXYLIC ACIDS BY HYDROLYZING NITRILES

Converting the Nitrile into a Carboxylic Acid

There are two ways of doing this, both of which involve reacting the carbon-nitrogen triple bond with water. This is described as hydrolysis.

The two methods produce slightly different products - you just have to be careful to get this right.

Acid Hydrolysis

The nitrile is heated under reflux with a dilute acid such as dilute hydrochloric acid. A carboxylic acid is formed. For example, starting from ethanenitrile you would get ethanoic acid. The ethanoic acid could be distilled off the mixture.

$$CH_3CN + 2H_2O + H^+ \longrightarrow CH_3COOH + NH_4^+$$

Alkaline Hydrolysis

The nitrile is heated under reflux with an alkali such as sodium hydroxide solution.

This time you wouldn't, of course, get a carboxylic acid produced - any acid formed would react with the sodium hydroxide present to give a salt. You also wouldn't get ammonium ions because they would react with sodium hydroxide to produce ammonia.

Starting from ethanenitrile, you would therefore get a solution containing ethanoate ions (for example, sodium ethanoate if you used sodium hydroxide solution) and ammonia.

$$CH_3CN + H_2O + OH^- \longrightarrow CH_3COO^- + NH_3$$

You have to remember to convert the ions into the free carboxylic acid, because that's what we are trying to make. To liberate the weak acid, ethanoic acid, you just have to supply hydrogen ions from a strong acid such as hydrochloric acid. You add enough hydrochloric acid to the mixture to make it acidic.

$$CH_3COO^- + H^+ \longrightarrow CH_3COOH$$

Now you can distill off the carboxylic acid.

PHYSICAL PROPERTIES

The physical properties (for example, boiling point and solubility) of the carboxylic acids are governed by their ability to form hydrogen bonds.

Boiling Points

Before we look at carboxylic acids, a reminder about alcohols:

The boiling points of alcohols are higher than those of alkanes of similar size because the alcohols can form hydrogen bonds with each other as well as Van der Waals dispersion forces and dipole-dipole interactions. The boiling points of carboxylic acids are some of the highest because of the

Hydrogen bonding involving both the Carbonyl Oxygen and the Hydroxyl Oxygen. Carboxylic acids have even higher Boiling Points than alcohols of the same number of carbons. The solubilities of carboxylic acids are very similar to the alcohols. Both families have members that are polar.

Carboxylic acids are weak acids giving up the carboxyl proton:

R-COOH $\longrightarrow$ $RCOO^- + H^+$ Since they are weak acids they form an equilibrium. The equilibrium constant is called K_a

$$K_a = [RCOO^-]\,[H^+] \,/\, [RCOOH]$$

The boiling points of carboxylic acids of similar size are higher still.

For example:

propan-1-ol	$CH_3CH_2CH_2OH$	97.2°C
ethanoic acid	CH_3COOH	118°C

These are chosen for comparison because they have identical relative molecular masses and almost the same number of electrons (which affects Van der Waals dispersion forces).

The higher boiling points of the carboxylic acids are still caused by hydrogen bonding, but operating in a different way.

In a pure carboxylic acid, hydrogen bonding can occur between two molecules of acid to produce a *dimer*.

This immediately doubles the size of the molecule and so increases the Van der Waals dispersion forces between one of these dimers and its neighbours - resulting in a high boiling point.

Solubility in Water

In the presence of water, the carboxylic acids don't dimerise. Instead, hydrogen bonds are formed between water molecules and individual molecules of acid.

The carboxylic acids with up to four carbon atoms will mix with water in any proportion. When you mix the two together, the energy released when the new hydrogen bonds form is much the same as is needed to break the hydrogen bonds in the pure liquids.

The solubility of the bigger acids decreases very rapidly with size. This is because the longer hydrocarbon "tails" of the molecules get between water molecules and break hydrogen bonds. In this case, these broken hydrogen bonds are only replaced by much weaker Van der Waals dispersion forces. The energetics of dissolving carboxylic acids in water is made more complicated because some of the acid molecules actually react with the water rather than just dissolving in it. This is the basis for the acidity of these compounds.

The Acidity of the Carboxylic Acids

Using the definition of an acid as a "substance which donates protons (hydrogen ions) to other things", the carboxylic acids are acidic because of the hydrogen in the -COOH group.

In solution in water, a hydrogen ion is transferred from the -COOH group to a water molecule. For example, with ethanoic acid, you get an ethanoate ion formed together with a hydroxonium ion, H_3O^+.

This reaction is reversible and, in the case of ethanoic acid, no more than about 1% of the acid has reacted to form ions at any one time.

These are therefore *weak acids.*

$$CH_3COOH + H_2O \rightleftharpoons CH_3COO^- + H_3O^+$$

This equation is often simplified to:

$$CH_3COOH_{(aq)} \rightleftharpoons CH_3COO^-_{(aq)} + H^+_{(aq)}$$

However, if you are going to use this second equation, you must include state symbols. They imply that the hydrogen ion is actually attached to a water molecule.

The pH of Carboxylic Acid Solutions

The pH depends on both the concentration of the acid and how easily it loses hydrogen ions from the -COOH group.

Ethanoic acid is typical of the acids where the -COOH group is attached to a simple alkyl group. Typical lab solutions have pH's in the 2 - 3 range, depending on their concentrations.

Methanoic acid is rather stronger than the other simple acids, and solutions have pH's about 0.5 pH units less than ethanoic acid of the same concentration.

REACTIONS

With Metals

Carboxylic acids react with the more reactive metals to produce a salt and hydrogen. The reactions are just the same as with acids like hydrochloric acid, except they tend to be rather slower.

For example, dilute ethanoic acid reacts with magnesium. The magnesium reacts to produce a colourless solution of magnesium ethanoate, and hydrogen is given off. If you use magnesium ribbon, the reaction is less vigorous than the same reaction with hydrochloric acid, but with magnesium powder, both are so fast that you probably wouldn't notice much difference.

$$2CH_3COOH + Mg \longrightarrow (CH_3COO)_2Mg + H_2$$

With Metal Hydroxides

These are simple neutralization reactions and are just the same as any other reaction in which hydrogen ions from an acid react with hydroxide ions. They are most quickly and easily represented by the equation:

$$H^+_{(aq)} + OH^-_{(aq)} \longrightarrow H_2O_{(l)}$$

If you mix dilute ethanoic acid with sodium hydroxide solution, for example, you simply get a colourless solution containing sodium ethanoate. The only sign that a change has happened is that the temperature of the mixture will have increased.

This change could well be represented by the ionic equation above, but if you want it, the full equation for this particular reaction is:

$$CH_3COOH + NaOH \longrightarrow CH_3COONa + H_2O$$

With Carbonates and Hydrogencarbonates

In both of these cases, a salt is formed together with carbon dioxide and water. Both are most easily represented by ionic equations.

For carbonates:

$$2H^+_{(aq)} + CO_3^{2-}{}_{(s)} \longrightarrow H_2O_{(l)} + CO_{2(g)}$$

and for hydrogencarbonates:

$$H^+_{(aq)} + HCO_{3(s)} \longrightarrow H_2O_{(l)} + CO_{2(g)}$$

If you pour some dilute ethanoic acid onto some white sodium carbonate or sodium hydrogencarbonate crystals, there is an immediate fizzing as carbon dioxide is produced. You end up with a colourless solution of sodium ethanoate.

With sodium carbonate, the full equation is:

$$2CH_3COOH + Na_2CO_3 \longrightarrow 2CH_3COONa + H_2O + CO_2$$

and for sodium hydrogencarbonate:

$$CH_3COOH + NaHCO_3 \longrightarrow CH_3COONa + H_2O + CO_2$$

There is very little obvious difference in the vigour of these reactions compared with the same reactions with dilute hydrochloric acid.

However, you would notice the difference if you used a slower reaction - for example with calcium carbonate in the form of a marble chip. With ethanoic acid, you would eventually produce a colourless solution of calcium ethanoate.

In this case, the marble chip would react noticeably more slowly with ethanoic acid than with hydrochloric acid.

With Ammonia

Ethanoic acid reacts with ammonia in exactly the same way as any other acid does. It transfers a hydrogen ion to the lone pair on the nitrogen of the ammonia and forms an ammonium ion.

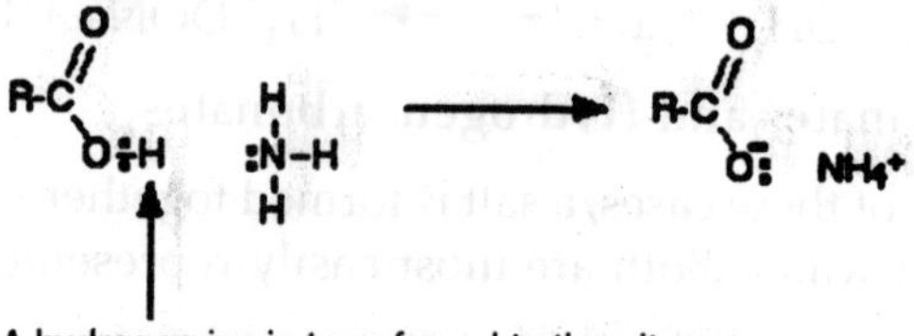

A hydrogen ion is transfemed to the nitrogen lone pair. Its electron is left behind on the oxygen making it negative

If you mix together a solution of ethanoic acid and a solution of ammonia, you will get a colourless solution of ammonium ethanoate.

$$CH_3COOH + NH_3 \longrightarrow CH_3COONH_4$$

With Amines

Amines are compounds in which one or more of the hydrogen atoms in an ammonia molecule have been replaced by a hydrocarbon group such as an alkyl group. For simplicity, we'll just look at compounds where only one of the hydrogen atoms has been replaced. These are called *primary amines.*

Simple primary amines include:

$CH_3\text{-}NH_2$	$CH_3\text{-}CH_2\text{-}NH_2$
methylamine (aminomethane)	ethylamine (or aminoethane)

The small amines are very similar indeed to ammonia in many ways. For example, they smell very much like ammonia and are just as soluble in water. Because all you have done to an ammonia molecule is swap a hydrogen for an alkyl group, the lone pair is still there on the nitrogen atom.

That means that they will react with acids (including carboxylic acids) in just the same way as ammonia does.

For example, ethanoic acid reacts with methylamine to produce a colourless solution of the salt methylammonium ethanoate.

$$CH_3COOH + CH_3NH_2 \longrightarrow CH_3COO^- \; CH_3NH_3^+$$

However complicated the amine, because all of them have got a lone pair on the nitrogen atom, you would get the same sort of reaction.

Esterification of Carboxylic Acids

Esters have a hydrocarbon group of some sort replacing the hydrogen in the -COOH group of a carboxylic acid.

Esters are produced when carboxylic acids are heated with alcohols in the presence of an acid catalyst. The catalyst is usually concentrated sulphuric acid. Dry hydrogen chloride gas is used in some cases, but these tend to involve aromatic esters (ones containing a benzene ring). The esterification reaction is both slow and reversible. The equation for the reaction between an acid RCOOH and an alcohol R'OH (where R and R' can be the same or different) is:

$$R\text{-}C(=O)\text{-}OH + R'OH \rightleftharpoons R\text{-}C(=O)\text{-}O\text{-}R' + H_2O$$

So, for example, if you were making ethyl ethanoate from ethanoic acid and ethanol, the equation would be:

$$CH_3\text{-}C(=O)\text{-}OH + CH_3CH_2OH \rightleftharpoons CH_3\text{-}C(=O)\text{-}O\text{-}CH_2CH_3 + H_2O$$

Doing the Reactions

On a Test Tube Scale

Carboxylic acids and alcohols are often warmed together in the presence of a few drops of concentrated sulphuric acid in order to observe the smell of the esters formed.

You would normally use small quantities of everything heated in a test tube stood in a hot water bath for a couple of minutes.

Because the reactions are slow and reversible, you don't get a lot of ester produced in this time. The smell is often masked or distorted by the smell of the carboxylic acid. A simple way of detecting the smell of the ester is to pour the mixture into some water in a small beaker.

Esters are virtually insoluble in water and tend to form a thin layer on the surface. Excess acid and alcohol both dissolve and are tucked safely away under the ester layer.

Small esters like ethyl ethanoate smell like typical organic solvents (ethyl ethanoate is a common solvent in, for example, glues).

As the esters get bigger, the smells tend towards artificial fruit flavouring - "pear drops", for example.

On a Larger Scale

If you want to make a reasonably large sample of an ester, the method used depends to some extent on the size of the ester. Small esters are formed faster than bigger ones.

To make a small ester like ethyl ethanoate, you can gently heat a mixture of ethanoic acid and ethanol in the presence of concentrated sulphuric acid, and distil off the ester as soon as it is formed.

This prevents the reverse reaction happening. It works well because the ester has the lowest boiling point of anything present. The ester is the only thing in the mixture which doesn't form hydrogen bonds, and so it has the weakest intermolecular forces.

Larger esters tend to form more slowly. In these cases, it may be necessary to heat the reaction mixture under reflux for some time to produce an equilibrium mixture. The ester can be separated from the carboxylic acid, alcohol, water and sulphuric acid in the mixture by fractional distillation.

REDUCTION

Here we are going to discuss the reduction of carboxylic acids to primary alcohols using lithium tetrahydridoaluminate(III) (lithium aluminium hydride), $LiAlH_4$.

The "(III)" is the oxidation state of the aluminium. Since aluminium only ever shows the +3 oxidation state in its compounds, the "(III)" is actually unnecessary.

The Reducing Agent

Lithium tetrahydridoaluminate has the structure:

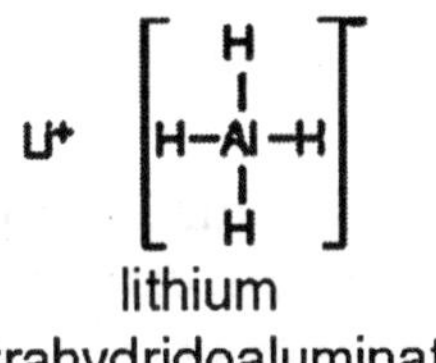

lithium tetrahydridoaluminate

In the negative ion, one of the bonds is a co-ordinate covalent (dative covalent) bond using the lone pair on a hydride ion (H^-) to form a bond with an empty orbital on the aluminium.

The Reduction of a Carboxylic Acid

The reaction happens in two stages - first to form an aldehyde and then a primary alcohol. Because lithium tetrahydridoaluminate reacts rapidly with aldehydes, it is impossible to stop at the halfway stage.

Equations for these reactions are usually written in a simplified form. The "[H]" in the equations represents hydrogen from a reducing agent.

Because of the impossibility of stopping at the aldehyde, there isn't much point in giving an equation for the two separate stages. The overall reaction is:

$$RCOOH + 4(H) \longrightarrow RCH_2OH + H_2O$$

"R" is hydrogen or a hydrocarbon group. For example, ethanoic acid will reduce to the primary alcohol, ethanol.

$$CH_3COOH + 4(H) \longrightarrow CH_3CH_2OH + H_2o$$

Sodium Tetrahydridoborate (Sodium Borohydride)

If you are familiar with the reduction of aldehydes and ketones using lithium tetrahydridoaluminate, you are probably aware that sodium tetrahydridoborate is often used as a safer alternative.

It CAN'T be used with carboxylic acids. The sodium tetrahydridoborate isn't reactive enough to reduce carboxylic acids.

Reaction Conditions

Lithium tetrahydridoaluminate reacts violently with water and so the reactions are carried out in solution in dry ethoxyethane (diethyl ether or just "ether"). The reaction happens at room temperature.

At the end of the reaction, the product is a complex aluminium salt. This is converted into the alcohol by treatment with dilute sulphuric acid.

$$\left(R\text{-}CH_2\text{-}O\text{-}\right)_4 Al^- + 4H^+_{(aq)} \longrightarrow 4\,R\text{-}CH_2\text{-}O\text{-}H + Al^{3+}_{(aq)}$$

DECARBOXYLATION OF CARBOXYLIC ACIDS AND THEIR SALTS

Decarboxylation using Soda Lime

A carboxylic acid has the formula RCOOH where R can be hydrogen or a hydrocarbon group such as an alkyl group. The hydrocarbon group could equally well be based on a benzene ring.

The sodium salt of a carboxylic acid will have the formula RCOONa.

In decarboxylation, the -COOH or -COONa group is removed and replaced with a hydrogen atom.

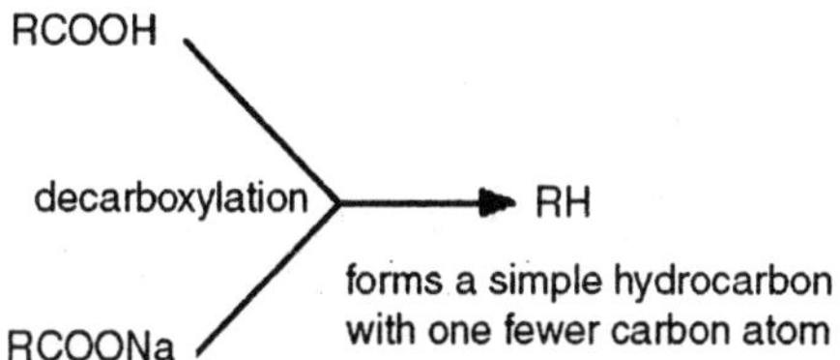

Soda lime is manufactured by adding sodium hydroxide solution to solid calcium oxide (quicklime). It is essentially a mixture of sodium hydroxide, calcium oxide and calcium hydroxide. It comes as white granules.

In equations, it is almost always written as if it were simply sodium hydroxide.

It is an easier material to handle than solid sodium hydroxide. Solid sodium hydroxide absorbs water from the atmosphere and you tend to end up with puddles of extremely concentrated (and corrosive) sodium hydroxide solution if you leave it exposed to the air.

Soda lime has much less tendency to absorb water.

The Reaction

The solid sodium salt of a carboxylic acid is mixed with solid soda lime, and the mixture is heated.

For example, if you heat sodium ethanoate with soda lime, you get methane gas formed:

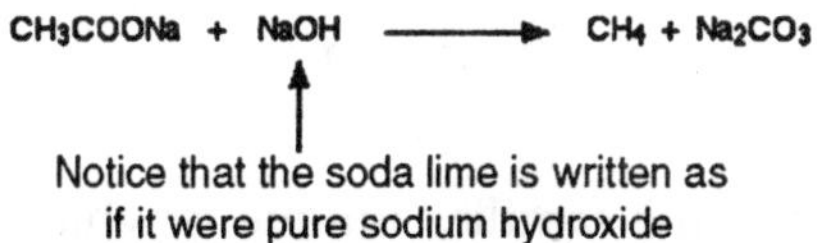

This reaction can be done with certain carboxylic acids themselves, but the acid would have to be solid. For example, benzene can be made by heating soda lime with solid benzoic acid (benzenecarboxylic acid), C6H5COOH.

$$C_6H_5COOH + 2NaOH \longrightarrow C_6H_6 + Na_2CO_3 + H_2O$$

You can think of this as first a reaction between the acid and the soda lime to make sodium benzoate, and then a decarboxylation as in the first example.

Chapter 13

Derivatives of Carboxylic Acids

ACYL HALIDE

Introduction

The Acyl Halide Family is a family of organic compounds with the functional group being the -COX. This group is attached to one of the carbons in the rest of the molecule. It is actually a carbonyl group, C=O, bonded to a Halogen atom. The halogen atom is an excellent leaving group so the main type of reaction for acyl halides is nucleophilic substitutions. Of all the acyl families this one is the most reactive. Acyl Families are families that have the acyl group as a part of their structure. The acyl group is:

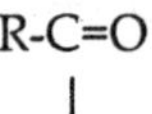

What makes the families unique is what atom or group of atoms are attached to the acyl carbon which is the carbon double bonded to the Oxygen. In a nucleophilic reaction that group is the potential leaving group that would be displaced during a nucleophilic substitution. For our discussion here we will symbolize the leaving group generally as "Y". So we can depict an acyl family of compounds as follows:

```
R-C=O
  |
  Y
```

1. If Y = –H, the compound belongs to the aldehyde family.
2. If Y = –R a hydrocarbon, the compound belongs to the ketone family
3. If Y = $-NR_2$ where the R's bonded to the Nitrogen can be either a H or a hydrocarbon, the compound belongs to the amide family.
4. If Y = –OR where R would be a hydrocarbon, the compound would belong to the ester family
5. If Y = –OH, the compound would belong to the Carboxylic Acid family
6. If Y = –OOCR, the compound would come from an acyl anhydride family
7. If Y = X where X could be any halogen atom, the compound would be an acyl halide

Acyl chlorides (also known as acid chlorides) are an example of an acid derivative. In this case, the -OH group has been replaced by a chlorine atom. For example:

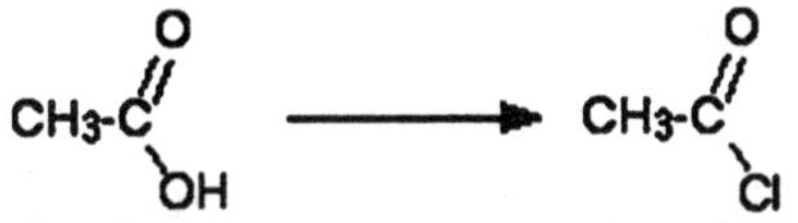

$CH_3-C(=O)-OH \longrightarrow CH_3-C(=O)-Cl$

ethanoic acid → ethanoyl chloride

PREPARATION OF ACYL HALIDES

Acyl Halides cannot be prepared from one of the members of the other acyl families because they represent the most reactive of the families. As soon as the acyl halide is produced it would be more reactive than what formed it and would reverse its own preparation. However we can form acyl halides by halogenating a carboxylic acid. The halogenating agent can be PX_3, PX_5, or SOX_2 where the X can be any halide:

$$CH_3COOH + SOCl_2 \rightarrow CH_3COCl + SO_2(g) + HCl(g)$$

or

$$CH_3COOH + PBr_5 \rightarrow CH_3COBr + HOPBr_4$$

or

$$3CH_3COOH + PCl_3 \longrightarrow 3CH_3COCl + H_3PO_3$$

Two kinds of reactions involving nucleophiles and acyl compounds are:

1. Nucleophilic Additions

```
R-C=O + Nuc⁻ H⁺ → R-C-O-H
  |                 |
  Y                 Y
```

2. Nucleophilic Substitutions

```
R-C=O  + Nuc⁻ → R-C=O + Y⁻
  |               |
  Y               Nuc
```

The factor that decides whether the acyl compound will undergo addition or substitution largely depends upon the ability for the "Y" group attached to the acyl carbon to be displaced by the incoming nucleophile. Different atoms or groups of atoms will have different abilities to leave the acyl compound or be displaced. The greater the "leaving group ability"(LGA) the more likely the acyl compound will undergo substitution instead of addition.

It might be instructive for us to review the ability for groups to leave an acyl carbon. Leaving group ability is largely dependent upon how well the group is stabilized after leaving the molecule. The more stable the group is the better able it is to depart and the more it will be involved in a substitution reaction with an incoming nucleophile. This stability can be associated with its bascicity. Weaker bases will be so stable that they won't want to donate electrons. So we can say that the weaker the basic strength of the leaving group the less likely for it to reverse the substitution step and the greater its ability to be displaced. We have also learned in previous lessons that weak bases come from strong acids forming them as a conjugate base.

For example, we all know that HCl is a relatively strong acid, so when it donates a proton it produces the conjugate base, Chloride ion. The very fact that HCl is such a strong acid will mean that Chloride ion has to be a weak conjugate base of that acid and therefore has little inclination to reverse the acid base reaction. For all practical purposes, the reaction of HCl with say water as the base will be one sided directed virtually exclusively to the right because Chloride ion is not basic enough to pull a proton away from the Hydronium ion that has formed:

$$H\text{-}Cl + H_2O \rightarrow H_3O^+ + Cl^-$$

Generally speaking, stronger acids and bases always give rise to weaker acids and bases. Therefore the stronger the acid the weaker its conjugate base. Weak conjugate bases (coming from strong acids) always make better leaving groups. These groups attached to an acyl carbon will always stimulate nucleophilic substitution over addition.

Relative strength of acids are as follows:

H-X is a stronger acid than carboxylic acids which are stronger than water H_2O which is stronger than alcohols R-OH which are stronger acids than R-NR_2(amines) which are much stronger acids than H_2 which is stronger as an acid than hydrocarbons.

Therefore, the relative strengths of their conjugate bases would be the reverse as follows:

R^- are stronger bases than H^- which are stronger than NR_2^- which are stronger bases than R-O^- which are stronger than OH^- which is stronger base than $RCOO^-$ which are stronger than X^-.

This means that the relative leaving group ability is as follows:

X^- greater than $RCOO^-$ which is greater than OH^- which are greater than R-O^- which are greater than NR_2^- which are much greater than H^- which are greater than R^-

In fact, Hydride ions and carbanions are so lousy as leaving groups substitutions rarely ,if ever, occur when they are attached to an acyl carbon. We are talking aldehydes and ketones here. Since substitutions are unlikely because of the relatively poor LGA(basicity) then nucleophilic additions are usually take place between such acyl compounds and an incoming nucleophile.

The reactivity toward nucleophilic substitutions for the rest of the acyl families would be:

Acyl halides are more reactive than acyl anhydrides which in turn are more reactive than carboxylic acids which are more reactive than esters which are more reactive than amides.

HYDROLYSIS OF ACYL HALIDES

Acyl halides will react with water as a nucleophile to form the Carboxylic acid and a strong Hydrogen Halide whish will react with water itself to form a corrosive strong acid which can be very irritating to the eyes and skin. These substances are called lacrymators. They stimulate the tear ducts and are used in tear gas used by the law enforcement agencies. The Hydrogen halide is formed from the hydrolysis of the acyl halide. The HX will then dissolve in the water of the eyes or the moisture in the skin and produce a highly corrosive and very irritating strong acid. If an acyl Chloride is used that would be Hydrochloric Acid. Acyl Bromides would produce Hydrobromic Acid. Here is an example of Hydrolysis:

1. Net Reaction: $CH_3COCl + H_2O \rightarrow CH_3COOH + HCl(g)$
2. Reaction Mechanism:

The reaction mechanism involves the nucleophilic attack of the water molecule as it attaches itself by way of the lone pair on the Oxygen (step 1).This causes the Oxygen to be positively charged. The resulting bond on the acyl Carbon will displace the Pi electrons out onto the acyl Oxygen(step 2). These electrons in turn come back reforming the Pi Bond of the acyl group and displace the Chlorine as a Chloride ion (step 3). A water molecule will pull off a proton from the Oxygen that bonded to the acyl Carbon in the first step producing a Hydronium ion(step 4). It is the Hydronium ion that does all the damage.

ALCOHOLYSIS OF ACYL HALIDES TO CARBOXYLIC ESTERS

The reaction of a Acyl Halide with an alcohol is an alcoholysis reaction which results in an ester being formed. This is also called an esterification. The reaction is relatively fast and requires no catalyst. This is because the leaving group in this substitution is a Chloride ion which is a relatively weak base and therefore an excellent leaving group. This is a good way to prepare an ester better than using a Carboxylic Acid.

Net Reaction:

$RCOCl + R'OH \rightarrow RCOOR' + HCl$

The mecahnism has the alcohol molecule attacking as a nucleophile the Oxygen attaching itself to the acyl Carbon displacing the Pi electrons out on the acyl Oxygen(step 1). The Pi electrons reform the Pi Bond thereby displacing the Cl group of the carboxyl group as a Chloride ion(step 2). The Chloride ion removes the Hydrogen on the alcohol to produce the ester(step 3).

AMMONOLYSIS OF ACYL HALIDE TO AMIDES

Ammonolysis is the use of a Nitrogen containing compound as the neucleophile which displaces the Cl group of the carboxyl group of the acid. The product is an amide. The reaction goes very well because of the high Leaving Group Ability of the Chlorine. If ammonia, NH_3 is used the amide is a simple amide. If a primary amine is used as the nucleophile then a mono-substituted amide is produced. If a secondary amine is used then the product is a di-substituted amide.

1. Use of Ammonia

   $RCOCl + 2NH_3 \rightarrow RCONH_2 + NH_4Cl$

2. Use of a primary amine

   $RCOCl + 2R'NH_2 \rightarrow RCONHR' + R'NH_3Cl$

3. Use of a secondary amine

   $RCOCl + 2R'R''NH \rightarrow RCONR'R'' + R'R''NH2Cl$

Mechanism:

The mechanism as is the case with all nucleophilic substitutions is similar to what happens in alcoholysis and hydrolysis. The nitrogen's lone pair is used to bond to the acyl carbon thereby displacing the Pi electrons of the acyl group onto the acyl Oxygen(step 1).The Pi electrons then reform the Pi Bond effectively displacing the Chlorine atom as a chloride ion(step 2).A second nitrogen containing molecule is basic enough to pull a Hydrogen off of the first nitrogen attached to the acyl carbon in step 1. This forms an ammonium salt with the Chloride ion(step 3). That is why it takes twice as much of the Nitrogen containing nucleophile in the reaction since one molecule is used in step 1 and a second molecule of the nucleophile is used as a base in step 3.

REACTION OF ACYL CHLORIDES WITH BENZENE

Friedel-Crafts acylation of Benzene

Acylation is the term given to substituting an acyl group such as CH_3CO- into another molecule. An acyl group is a hydrocarbon group attached to a carbon-oxygen double bond.

The most commonly used example of an acyl group is the ethanoyl group, CH_3CO-, and so that's the one we will stick with throughout.

So, if you react benzene with ethanoyl chloride in the presence of an aluminium chloride catalyst, the equation for the reaction is:

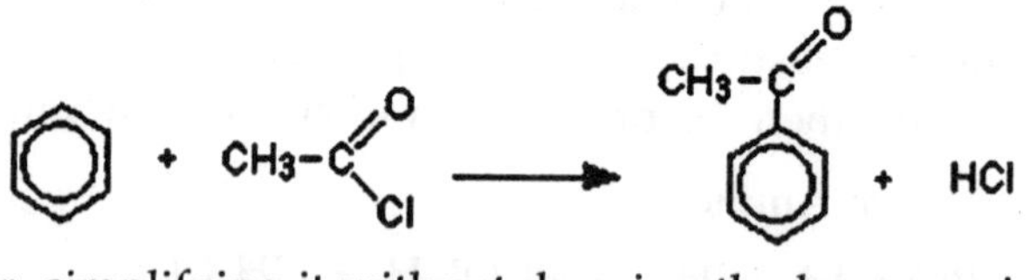

Or, simplifying it without drawing the benzene ring:

$$C_6H_6 + CH_3COCl \longrightarrow C_6H_5COCH_3 + HCl$$

In the simplified formula for the product, the phenyl group is usually written on the left-hand side and the alkyl group to the right of the carbon-oxygen double bond.

The aluminium chloride isn't written into these equations because it is acting as a catalyst. If you wanted to include it, you could write $AlCl_3$ over the top of the arrow.

The product is called phenylethanone (old name, acetophenone).

Doing the Reaction

Ethanoyl chloride is added carefully to a mixture of benzene and solid aluminium chloride in the cold. Hydrogen chloride gas is given off.

When all the ethanoyl chloride has been added, the mixture is heated under reflux at a temperature of 60°C for about 30 minutes to complete the reaction.

Why the Reaction is Important

Friedel-Crafts acylation is a very effective way of attaching a hydrocarbon-based group to a benzene ring. Although the product is a ketone (a compound containing a carbon-oxygen double bond with a hydrocarbon group either side), it is easily converted into other things.

For example:

The carbon-oxygen double bond can be reduced to give a secondary alcohol, which in turn can undergo a whole lot of other reactions.

Phenylethanone can also be reduced to produce ethylbenzene.

This is known as the Clemmensen reduction and involves heating the ketone with amalgamated zinc (a mixture of zinc and mercury) and concentrated hydrochloric acid for a long time.

$$C_6H_5COCH_3 \xrightarrow[\text{COnc HCl}]{\text{Zn/Hg}} C_6H_5CH_2CH_3$$

This indirect route is the best way of getting an alkyl group attached to a benzene ring.

It is possible to attach an alkyl group directly to the ring, but it is impossible to stop at substituting just one. An alkyl group attached to the ring makes the ring more reactive than the original benzene. That means that something like ethylbenzene reacts faster than benzene itself.

The result is that you get several ethyl groups substituted around the ring rather than just one.

Attaching an acyl group to the ring makes the ring so unreactive that it won't substitute a second one.

ESTERS

Introduction

Esters are derived from carboxylic acids. A carboxylic acid contains the -COOH group, and in an ester the hydrogen in this group is replaced by a hydrocarbon group of some kind. This could be an alkyl group like methyl or ethyl, or one containing a benzene ring like phenyl.

A common ester - ethyl ethanoate is the most commonly discussed ester is ethyl ethanoate. In this case, the hydrogen in the -COOH group has been replaced by an ethyl group. The formula for ethyl ethanoate is:

$$CH_3\text{-}C(=O)\text{-}O\text{-}CH_2\text{-}CH_3$$

ethyl ethanoate

Notice that the ester is named the opposite way around from the way the formula is written. The "ethanoate" bit comes from ethanoic acid. The "ethyl" bit comes from the ethyl group on the end.

A few more Esters

In each case, be sure that you can see how the names and formulae relate to each other.

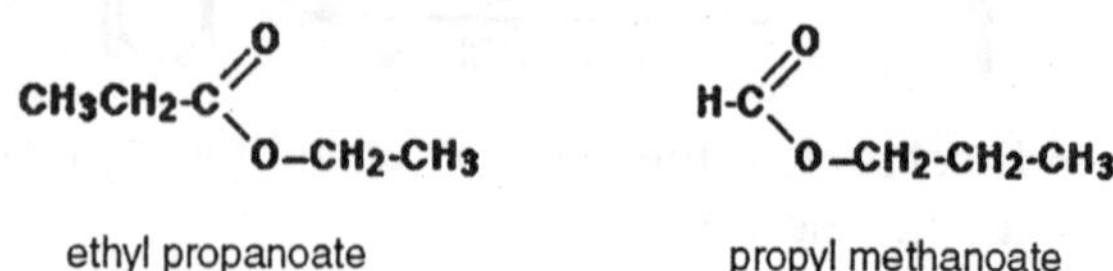

ethyl propanoate

propyl methanoate

Notice that the acid is named by counting up the total number of carbon atoms in the chain - *including the one in the -COOH group*. So, for example, CH_3CH_2COOH is propanoic acid, and CH_3CH_2COO is the propanoate group.

PREPARATION OF ESTERS

Making Esters using Carboxylic Acids

This method can be used for converting alcohols into esters, but it doesn't work with phenols - compounds where the -OH group is attached directly to a benzene ring. Phenols react with carboxylic acids so slowly that the reaction is unusable for preparation purposes.

The Chemistry of the Reaction

Esters are produced when carboxylic acids are heated with alcohols in the presence of an acid catalyst. The catalyst is usually concentrated sulphuric acid. Dry hydrogen chloride gas is used in some cases, but these tend to involve aromatic esters (ones where the carboxylic acid contains a benzene ring). The esterification reaction is both slow and reversible. The equation for the reaction between an acid RCOOH and an alcohol R'OH (where R and R' can be the same or different) is:

$$R\text{-}C(=O)\text{-}OH + R'OH \rightleftharpoons R\text{-}C(=O)\text{-}O\text{-}R' + H_2O$$

So, for example, if you were making ethyl ethanoate from ethanoic acid and ethanol, the equation would be:

$$CH_3\text{-}C(=O)\text{-}OH + CH_3CH_2OH \rightleftharpoons CH_3\text{-}C(=O)\text{-}O\text{-}CH_2CH_3 + H_2O$$

On a test Tube Scale

Carboxylic acids and alcohols are often warmed together in the presence of a few drops of concentrated sulphuric acid in order to observe the smell of the esters formed.

You would normally use small quantities of everything heated in a test tube stood in a hot water bath for a couple of minutes.

Because the reactions are slow and reversible, you don't get a lot of ester produced in this time. The smell is often masked or distorted by the smell of the carboxylic acid. A simple way of detecting the smell of the ester is to pour the mixture into some water in a small beaker.

Apart from the very small ones, esters are fairly insoluble in water and tend to form a thin layer on the surface. Excess acid and alcohol both dissolve and are tucked safely away under the ester layer.

Small esters like ethyl ethanoate smell like typical organic solvents (ethyl ethanoate is a common solvent in, for example, glues).

As the esters get bigger, the smells tend towards artificial fruit flavouring - "pear drops", for example.

On a Larger Scale

If you want to make a reasonably large sample of an ester, the method used depends to some extent on the size of the ester. Small esters are formed faster than bigger ones.

To make a small ester like ethyl ethanoate, you can gently heat a mixture of ethanoic acid and ethanol in the presence of concentrated sulphuric acid, and distil off the ester as soon as it is formed.

This prevents the reverse reaction happening. It works well because the ester has the lowest boiling point of anything present. The ester is the only thing in the mixture which doesn't form hydrogen bonds, and so it has the weakest intermolecular forces.

Making Esters using Acyl Chlorides (acid chlorides)

This method will work for alcohols and phenols. In the case of phenols, the reaction is sometimes improved by first converting the phenol into a more reactive form.

The Basic Reaction

If you add an acyl chloride to an alcohol, you get a vigorous (even violent) reaction at room temperature producing an ester and clouds of steamy acidic fumes of hydrogen chloride.

For example, if you add the liquid ethanoyl chloride to ethanol, you get a burst of hydrogen chloride produced together with the liquid ester ethyl ethanoate.

$$CH_3COCl + CH_3CH_2OH \longrightarrow CH_3COOCH_2CH_3 + HCl$$

The substance normally called "phenol" is the simplest of the family of phenols. Phenol has an -OH group attached to a benzene ring - and nothing else.

The reaction between ethanoyl chloride and phenol is similar to the ethanol reaction although not so vigorous. Phenyl ethanoate is formed together with hydrogen chloride gas.

$$CH_3\text{-}C(=O)Cl + C_6H_5OH \longrightarrow CH_3\text{-}C(=O)\text{-}O\text{-}C_6H_5 + HCl$$

Making Esters using Acid Anhydrides

This reaction can again be used to make esters from both alcohols and phenols. The reactions are slower than the corresponding reactions with acyl chlorides, and you usually need to warm the mixture.

In the case of a phenol, you can react the phenol with sodium hydroxide solution first, producing the more reactive phenoxide ion.

Taking ethanol reacting with ethanoic anhydride as a typical reaction involving an alcohol:

There is a slow reaction at room temperature (or faster on warming). There is no visible change in the colourless liquids, but a mixture of ethyl ethanoate and ethanoic acid is formed.

$$(CH_3CO)_2O + CH_3CH_2OH \longrightarrow CH_3COOCH_2CH_3 + CH_3COOH$$

The reaction with phenol is similar, but will be slower. Phenyl ethanoate is formed together with ethanoic acid.

$$(CH_3CO)_2O + C_6H_5OH \longrightarrow CH_3COOC_6H_5 + CH_3COOH$$

This reaction isn't important itself, but a very similar reaction is involved in the manufacture of aspirin.

If the phenol is first converted into sodium phenoxide by adding sodium hydroxide solution, the reaction is faster. Phenyl ethanoate is again formed, but this time the other product is sodium ethanoate rather than ethanoic acid.

$$(CH_3CO)_2O + C_6H_5O^-Na^+ \longrightarrow CH_3COOC_6H_5 + CH_3COO^-Na^+$$

Fats and Oils

Differences between Fats and Oils

Animal and vegetable fats and oils are just big complicated esters. The difference between a fat (like butter) and an oil (like sunflower oil) is simply in the melting points of the mixture of esters they contain.

If the melting points are below room temperature, it will be a liquid - an oil. If the melting points are above room temperature, it will be a solid - a fat.

The causes of the differences in melting points will be discussed further down the page under physical properties.

Fats and Oils as Big Esters

Esters can be made from carboxylic acids and alcohols. This is discussed in detail on another page, but in general terms, the two combine together losing a molecule of water in the process.

The diagram shows the relationship between the ethanoic acid, the ethanol and the ester.

$CH_3-C(=O)-O-H$ $HO-CH_2-CH_3$ → $CH_3-C(=O)-O-CH_2-CH_3$

Ethanol written back wards so that you can see the relationship better

ethyl ethanoate

This isn't intended to be a full equation. Water, of course, is also produced.

Now lets make the alcohol a bit more complicated by having more than one -OH group. The diagram below shows the structure of propane-1,2,3-triol (old name: glycerol).

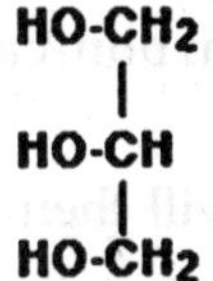

Just as with the ethanol in the previous equation, I've drawn this back-to-front to make the next diagrams clearer. Normally, it is drawn with the -OH groups on the right-hand side.

If you make an ester of this with ethanoic acid, you could attach three ethanoate groups.

CH_3COOCH_2
CH_3COOCH
CH_3COOCH_2

Now, make the acid chains much longer, and you finally have a fat.

$$
\begin{array}{l}
CH_3(CH_2)_{16}COOCH_2 \\
\qquad\qquad\qquad\quad | \\
CH_3(CH_2)_{16}COOCH \\
\qquad\qquad\qquad\quad | \\
CH_3(CH_2)_{16}COOCH_2
\end{array}
$$

The acid $CH_3(CH_2)_{16}COOH$ is called octadecanoic acid, but the old name is still commonly used. This is stearic acid.

The full name for the ester of this with propane-1,2,3-triol is propane-1,2,3-triyl trioctadecanoate. But the truth is that almost everybody calls it by its old name of glyceryl tristearate.

Saturated and Unsaturated Fats and Oils

If the fat or oil is saturated, it means that the acid that it was derived from has no carbon-carbon double bonds in its chain. Stearic acid is a saturated acid, and so glyceryl tristearate is a saturated fat.

If the acid has just one carbon-carbon double bond somewhere in the chain, it is called mono-unsaturated. If it has more than one carbon-carbon double bond, it is polyunsaturated.

Those same terms will then apply to the esters that are formed.

All of these are saturated acids, and so will form saturated fats and oils:

$CH_3(CH_2)_{10}COOH$ launc acid (dodecandic acid)

$CH_3(CH_2)_{14}COOH$ palmitic acid (hexadecanoic acid)

$CH_3(CH_2)_{16}COOH$ stearic acid (octadecanoic acid)

Oleic acid is a typical mono-unsaturated acid:

$$CH_3(CH_2)_7CH{=}CH(CH_2)_7COOH$$

oleic acid (octadec-9-enoic acid)

and linoleic and linolenic acids are typical polyunsaturated acids.

$CH_3(CH_2)_4CH{=}CHCH_2CH{=}CH(CH_2)_7COOH$

linoleic acid (Octadec-9, 12-dienoic acid)

$CH_3CH_2CH{=}CHCH_2CH{=}CHCH_2CH{=}CH(CH_2)_7COOH$

linolenic acid (octadec-9,12,15-trienoic acid)

You might possibly have come across the terms "omega 6" and "omega 3" in the context of fats and oils.

Linoleic acid is an omega 6 acid. It just means that the first carbon-carbon double bond starts on the sixth carbon from the CH_3 end.

Linolenic acid is an omega 3 acid for the same reason.

Because of their relationship with fats and oils, all of the acids above are sometimes described as fatty acids.

PHYSICAL PROPERTIES

Simple Esters Like Ethyl Ethanoate etc.

Boiling Points

The small esters have boiling points which are similar to those of aldehydes and ketones with the same number of carbon atoms.

Like aldehydes and ketones, they are polar molecules and so have dipole-dipole interactions as well as Van der Waals dispersion forces. However, they don't form hydrogen bonds, and so their boiling points aren't anything like as high as an acid with the same number of carbon atoms.

For example:

molecule	type	boiling point (°C)
$CH_3COOCH_2CH_3$	ester	77.1
$CH_3CH_2CH_2COOH$	carboxylic acid	164

Solubility in Water

The small esters are fairly soluble in water but solubility falls with chain length.

For example:

ester	formula	solubility (g per 100 g of water)
ethyl methanoate	$HCOOCH_2CH_3$	10.5
ethyl ethanoate	$CH_3COOCH_2CH_3$	8.7
ethyl propanoate	$CH_3CH_2COOCH_2CH_3$	1.7

The reason for the solubility is that although esters can't hydrogen bond with themselves, they *can* hydrogen bond with water molecules.

One of the slightly positive hydrogen atoms in a water molecule can be sufficiently attracted to one of the lone pairs on one of the oxygen atoms in an ester for a hydrogen bond to be formed.

There will also, of course, be dispersion forces and dipole-dipole attractions between the ester and the water molecules.

Forming these attractions releases energy. This helps to supply the energy needed to separate water molecule from water molecule and ester molecule from ester molecule before they can mix together.

As chain lengths increase, the hydrocarbon parts of the ester molecules start to get in the way.

By forcing themselves between water molecules, they break the relatively strong hydrogen bonds between water molecules without replacing them by anything as good. This makes the process energetically less profitable, and so solubility decreases.

THE PHYSICAL PROPERTIES OF FATS AND OILS

Solubility in Water

None of these molecules are water soluble. The chain lengths are now so great that far too many hydrogen bonds between water molecules would have to be broken - so it isn't energetically profitable.

Melting Points

The melting points determine whether the substance is a fat (a solid at room temperature) or an oil (a liquid at room temperature).

Fats normally contain saturated chains. These allow more effective Van der Waals dispersion forces between the molecules. That means you need more energy to separate them, and so increases the melting points.

The greater the extent of the unsaturation in the molecules, the lower the melting points tend to be because the Van der Waals dispersion forces are less effective.

Why should this be? We are talking about molecules of very similar sizes and so the potential for temporary dipoles should be much the same in all of them. What matters, though, is how close together the molecules can get.

Van der Waals dispersion forces need the molecules to be able to pack closely together to be really effective. The presence of carbon-carbon double bonds in the chains gets in the way of tidy packing.

Here is a simplified diagram of a saturated fat:

The hydrocarbon chains are, of course, in constant motion in the liquid, but it is possible for them to lie tidily when the substance solidifies. If the chains in one molecule can lie tidily, that means that neighbouring molecules can get close.

That increases the attractions between one molecule and its neighbours and so increases the melting point.

Unsaturated fats and oils have at least one carbon-carbon double bond in at least one chain.

There isn't any rotation about a carbon-carbon double bond and so that locks a permanent kink into the chain. That makes packing molecules close together more difficult. If they don't pack so well, the Van der Waals forces won't work as well.

This effect is much worse for molecules where the hydrocarbon chains either end of the double bond are arranged cis to each other - in other words, both of them on the same side of the double bond:

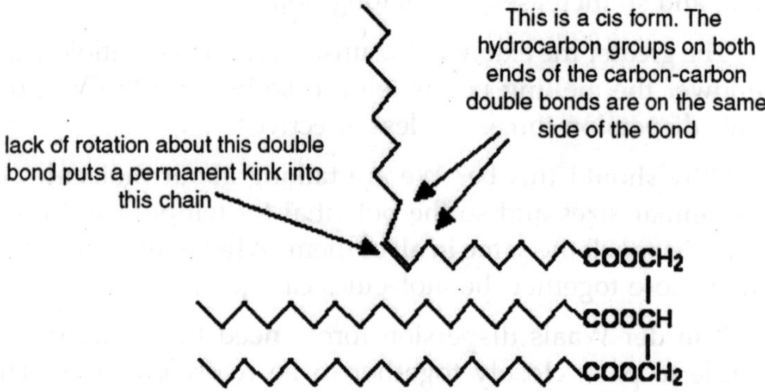

If they are on opposite sides of the double bond (the *trans* form) the effect isn't as marked. It is, however, rather more than the diagram below suggests because of the changes in bond angles around the double bond compared with the rest of the chain.

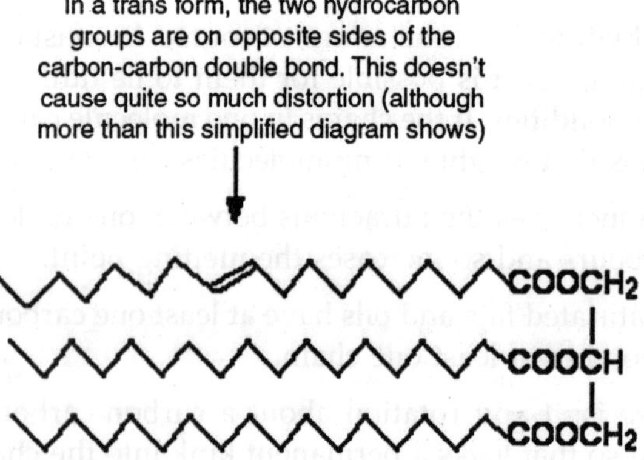

Trans fats and oils have higher melting points than cis ones because the packing isn't affected quite as much. Naturally occurring unsaturated fats and oils tend to be the cis form.

HYDROLYSIS

Hydrolyzing Simple Esters

Technically, hydrolysis is a reaction with water. That is exactly what happens when esters are hydrolyzed by water or by dilute acids such as dilute hydrochloric acid.

The alkaline hydrolysis of esters actually involves reaction with hydroxide ions, but the overall result is so similar that it is lumped together with the other two.

Hydrolysis using water or dilute acid

The reaction with pure water is so slow that it is never used. The reaction is catalysed by dilute acid, and so the ester is heated under reflux with a dilute acid like dilute hydrochloric acid or dilute sulphuric acid.

Here are two simple examples of hydrolysis using an acid catalyst.

First, hydrolyzing ethyl ethanoate:

$$\underset{\text{ethyl ethanoate}}{CH_3COOCH_2CH_3} + H_2O \xrightleftharpoons{H^+_{(aq)}} \underset{\text{ethanoic acid}}{CH_3COOH} + \underset{\text{ethanol}}{CH_3CH_2OH}$$

and then hydrolyzing methyl propanoate:

$$\underset{\text{methyl propanoate}}{CH_3CH_2COOCH_3} + H_2O \xrightleftharpoons{H^+_{(aq)}} \underset{\text{propanoic acid}}{CH_3CH_2COOH} + \underset{\text{methanol}}{CH_3OH}$$

Notice that the reactions are reversible. To make the hydrolysis as complete as possible, you would have to use an excess of water. The water comes from the dilute acid, and so you would mix the ester with an excess of dilute acid.

Hydrolysis using Dilute Alkali

This is the usual way of hydrolyzing esters. The ester is heated under reflux with a dilute alkali like sodium hydroxide solution.

There are two big adVantages of doing this rather than using a dilute acid. The reactions are one-way rather than reversible, and the products are easier to separate.

Taking the same esters as above, but using sodium hydroxide solution rather than a dilute acid:

First, hydrolyzing ethyl ethanoate using sodium hydroxide solution:

$$CH_3COOCH_2CH_3 + NaOH \longrightarrow CH_3COONa + CH_3CH_2OH$$

ethylethanoate — sodium ethanoate — ethanol

and then hydrolyzing methyl propanoate in the same way:

$$CH_3CH_2COOCH_3 + NaOH \longrightarrow CH_3CH_2COONa + CH_3OH$$

methyl propanoate — sodium propanoate — methanol

Notice that you get the sodium salt formed rather than the carboxylic acid itself.

This mixture is relatively easy to separate. Provided you use an excess of sodium hydroxide solution, there won't be any ester left - so you don't have to worry about that.

The alcohol formed can be distilled off.

If you want the acid rather than its salt, all you have to do is to add an excess of a strong acid like dilute hydrochloric acid or dilute sulphuric acid to the solution left after the first distillation.

If you do this, the mixture is flooded with hydrogen ions. These are picked up by the ethanoate ions (or propanoate ions or whatever) present in the salts to make ethanoic acid (or propanoic acid, etc). Because these are weak acids, once they combine with the hydrogen ions, they tend to stay combined.

The carboxylic acid can now be distilled off.

Hydrolyzing Complicated Esters to make soap

This next bit deals with the alkaline hydrolysis (using sodium hydroxide solution) of the big esters found in animal and vegetable fats and oils.

If the large esters present in animal or vegetable fats and oils are heated with concentrated sodium hydroxide solution exactly the same reaction happens as with the simple esters.

A salt of a carboxylic acid is formed - in this case, the sodium salt of a big acid such as octadecanoic acid (stearic acid). These salts are the important ingredients of soap - the ones that do the cleaning.

An alcohol is also produced - in this case, the more complicated alcohol, propane-1,2,3-triol (glycerol).

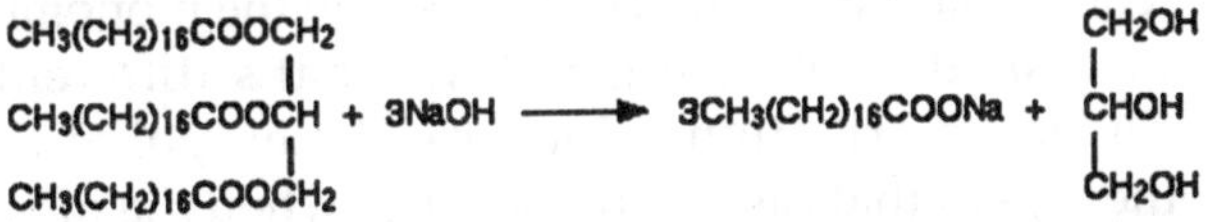

$$\begin{array}{l} CH_3(CH_2)_{16}COOCH_2 \\ \quad\quad\quad\quad\quad\quad | \\ CH_3(CH_2)_{16}COOCH \\ \quad\quad\quad\quad\quad\quad | \\ CH_3(CH_2)_{16}COOCH_2 \end{array} + 3NaOH \longrightarrow 3CH_3(CH_2)_{16}COONa + \begin{array}{l} CH_2OH \\ | \\ CHOH \\ | \\ CH_2OH \end{array}$$

a typical fat or oil — a typical sodium salt found in soap — glycerol

Because of its relationship with soap making, the alkaline hydrolysis of esters is sometimes known as *saponification*.

POLYESTERS

Here we look at the formation, structure and uses of a common polyester sometimes known as Terylene if it is used as a fibre, or PET if it used in, for example, plastic drinks bottles

Poly(Ethylene Terephthalate)

A polyester is a polymer (a chain of repeating units) where the individual units are held together by ester linkages.

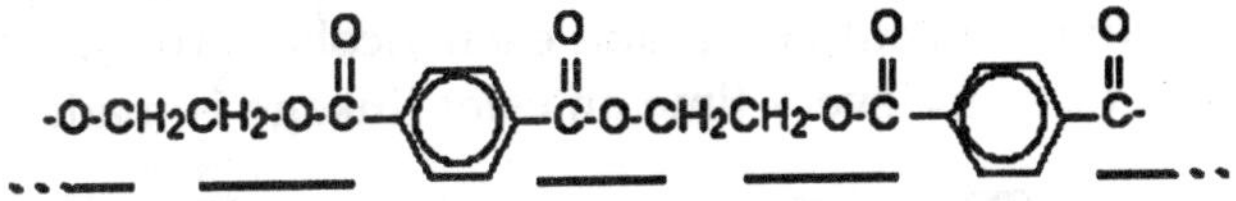

$$\ldots\text{-O-}CH_2CH_2\text{-O-}\overset{O}{\overset{\|}{C}}\text{-}C_6H_4\text{-}\overset{O}{\overset{\|}{C}}\text{-O-}CH_2CH_2\text{-O-}\overset{O}{\overset{\|}{C}}\text{-}C_6H_4\text{-}\overset{O}{\overset{\|}{C}}\text{-}\ldots$$

ester linkages

The diagram shows a very small bit of the polymer chain and looks pretty complicated. But it isn't very difficult to work out - and that's the best thing to do: work it out, not try to remember it. You will see how to do that in a moment.

The usual name of this common polyester is poly(ethylene terephthalate). The everyday name depends on whether it is being used as a fibre or as a material for making things like bottles for soft drinks.

When it is being used as a fibre to make clothes, it is often just called polyester. It may sometimes be known by a brand name like Terylene.

When it is being used to make bottles, for example, it is usually called PET.

Making Polyesters as an example of Condensation Polymerization

In condensation polymerization, when the monomers join together a small molecule gets lost. That's different from addition polymerization which produces polymers like poly(ethene) - in that case, nothing is lost when the monomers join together.

A polyester is made by a reaction involving an acid with two -COOH groups, and an alcohol with two -OH groups.

In the common polyester drawn above:

The acid is benzene-1,4-dicarboxylic acid (old name: terephthalic acid).

The alcohol is ethane-1,2-diol (old name: ethylene glycol).

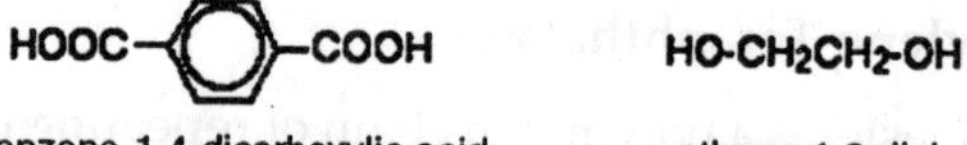

Now imagine lining these up alternately and making esters with each acid group and each alcohol group, losing a molecule of water every time an ester linkage is made.

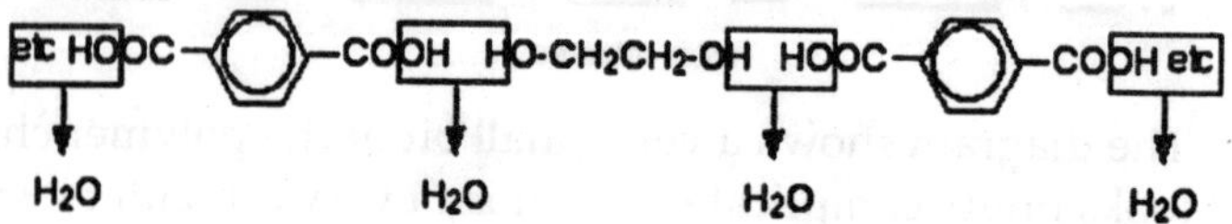

That would produce the chain shown above.

etc -OC–C_6H_4–$COOCH_2CH_2OOC$–C_6H_4–CO- etc

Manufacturing poly(ethylene terephthalate)

The reaction takes place in two main stages: a pre-polymerization stage and the actual polymerization.

In the first stage, before polymerization happens, you get a fairly simple ester formed between the acid and two molecules of ethane-1,2-diol.

$$HO\text{-}CH_2CH_2\text{-}OH \quad HOOC\text{-}C_6H_4\text{-}COOH \quad HO\text{-}CH_2CH_2\text{-}OH$$

$$\downarrow$$

$$HOCH_2CH_2OOC\text{-}C_6H_4\text{-}COOCH_2CH_2OH + 2H_2O$$

In the polymerization stage, this is heated to a temperature of about 260°C and at a low pressure. A catalyst is needed - there are several possibilities including antimony compounds like antimony(III) oxide.

The polyester forms and half of the ethane-1,2-diol is regenerated. This is removed and recycled.

$$n\ HOCH_2CH_2OOC\text{-}C_6H_4\text{-}COOCH_2CH_2OH$$

$$\downarrow$$

$$(\text{-}OC\text{-}C_6H_4\text{-}COOCH_2CH_2O\text{-})_n + n\ HOCH_2CH_2OH$$

Hydrolysis of Polyesters

Simple esters are easily hydrolyzed by reaction with dilute acids or alkalis.

Polyesters are attacked readily by alkalis, but much more slowly by dilute acids. Hydrolysis by water alone is so slow as to be completely unimportant. If you spill dilute alkali on a fabric made from polyester, the ester linkages are broken. Ethane-1,2-diol is formed together with the salt of the carboxylic acid.

Because you produce small molecules rather than the original polymer, the fibres are destroyed, and you end up with a hole.

For example, if you react the polyester with sodium hydroxide solution:

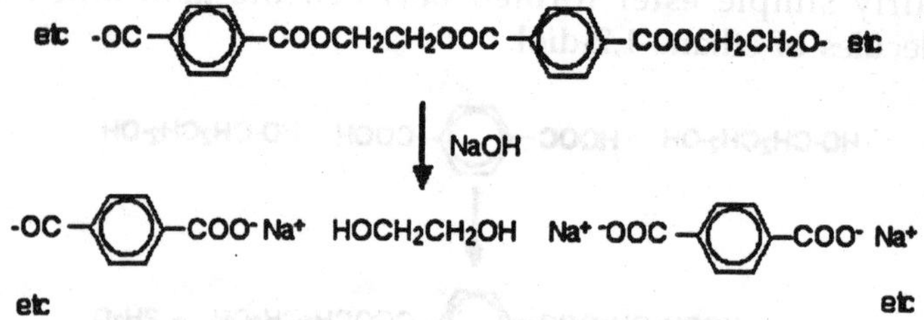

etc -OC–C_6H_4–$COOCH_2CH_2OOC$–C_6H_4–$COOCH_2CH_2O$- etc

↓ NaOH

-OC–C_6H_4–COO^-Na^+ $HOCH_2CH_2OH$ $Na^{+-}OOC$–C_6H_4–COO^-Na^+

etc   etc

ACID ANHYDRIDES

Introduction

A carboxylic acid such as ethanoic acid has the structure:

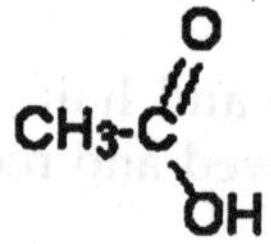

CH_3-C(=O)OH

ethanoic acid

If you took two ethanoic acid molecules and removed a molecule of water between them you would get the acid anhydride, ethanoic anhydride (old name: acetic anhydride).

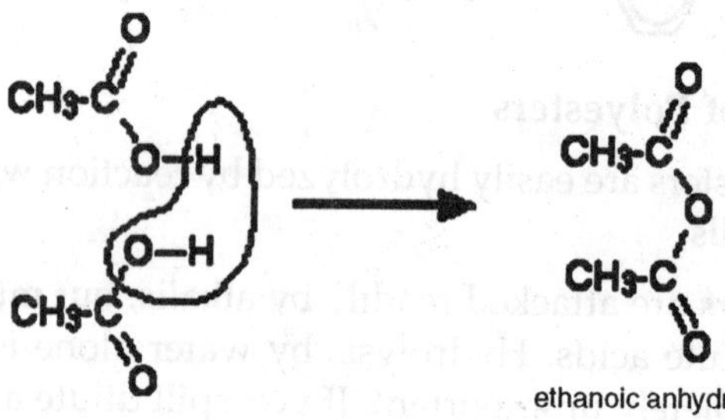

CH_3-C(=O)-O-H + H-O-C(=O)-CH_3 → CH_3-C(=O)-O-C(=O)-CH_3

ethanoic anhydride

You can actually make ethanoic anhydride by dehydrating ethanoic acid, but it is normally made in a more efficient, round-about way.

NAMING ACID ANHYDRIDES

You just need to take the name of the parent acid, and replace the word "acid" by "anhydride". "Anhydride" simply means "without water".

So ethanoic acid forms ethanoic anhydride; propanoic acid forms propanoic anhydride, and so on.

PHYSICAL PROPERTIES OF ACID ANHYDRIDES

We will take ethanoic anhydride as typical.

Appearance

Ethanoic anhydride is a colourless liquid, smelling strongly of vinegar (ethanoic acid).

The smell is because ethanoic anhydride reacts with water vapour in the air (and moisture in your nose) to produce ethanoic acid again.

Solubility in Water

Ethanoic anhydride can't be said to dissolve in water because it reacts with it to give ethanoic acid. There is no such thing as an aqueous solution of ethanoic anhydride.

Boiling Point

Ethanoic anhydride boils at 140°C. This is because it is a fairly big polar molecule and so has both Van der Waals dispersion forces and dipole-dipole attractions.

It doesn't, however, form hydrogen bonds. That means that its boiling point isn't as high as a carboxylic acid of similar size. For example, pentanoic acid (the most similarly sized acid) boils at 186°C.

REACTIVITY OF ACID ANHYDRIDES

Comparing acid anhydrides with acyl chlorides (acid chlorides)

You have almost certainly come across acid anhydrides for the first time just after looking at acyl chlorides, or you may be studying them at the same time as acyl chlorides.

It is much, much easier to think of acid anhydrides as if they were a sort of modified acyl chloride than to try to learn about them from scratch. That is the line I intend to take throughout all this section.

Compare the structure of an acid anhydride with that of an acyl chloride - looking carefully at the way it is colour-coded in the diagram.

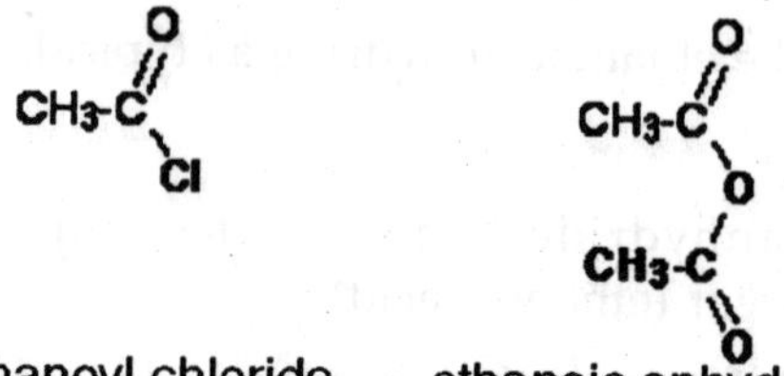

$$CH_3-C(=O)-Cl \qquad CH_3-C(=O)-O-C(=O)-CH_3$$

ethanoyl chloride ethanoic anhydride

In the reactions of ethanoic anhydride, the red group at the bottom always stays intact. It is behaving in many ways as if it was a single atom - just like the chlorine atom in the acyl chloride.

The usual reaction of an acyl chloride is replacement of the chlorine by something else.

Taking ethanoyl chloride as typical, the initial reaction is of this kind:

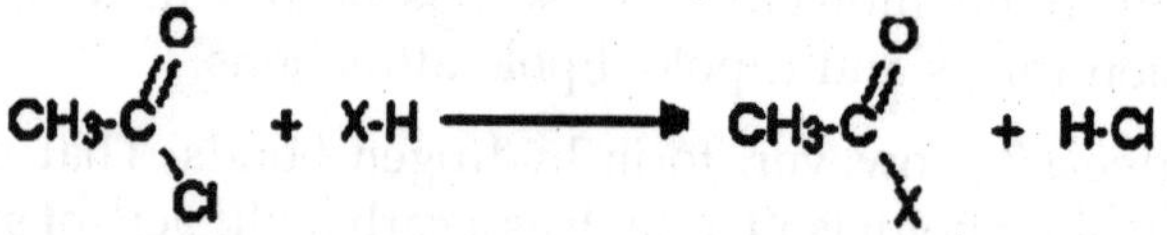

$$CH_3-C(=O)-Cl + X\text{-}H \longrightarrow CH_3-C(=O)-X + H\text{-}Cl$$

Hydrogen chloride gas is given off, although that might go on to react with other components of the mixture.

With an acid anhydride, the reaction is slower, but the only essential difference is that instead of hydrogen chloride being produced as the other product, you get ethanoic acid instead.

$$CH_3C(=O)\text{-}O\text{-}C(=O)CH_3 + X\text{-}H \longrightarrow CH_3C(=O)\text{-}X + CH_3COOH$$

Just like the hydrogen chloride, this might afterwards go on to react with other things present.

The reactions (of both acyl chlorides and acid anhydrides) involve things like water, alcohols and phenols, or ammonia and amines. All of these particular cases contain a very electronegative element with an active lone pair of electrons - either oxygen or nitrogen.

REACTION OF ACID ANHYDRIDES WITH WATER, ALCOHOLS AND PHENOL

Similarities between the Reactions

Comparing the Structures of Water, Ethanol and Phenol

Each substance contains an -OH group. In water, this is attached to a hydrogen atom. In an alcohol, it is attached to an alkyl group - shown in the diagrams below as "R". In phenols, it is attached to a benzene ring. Phenol (the simplest member of the family of phenols) is C_6H_5OH.

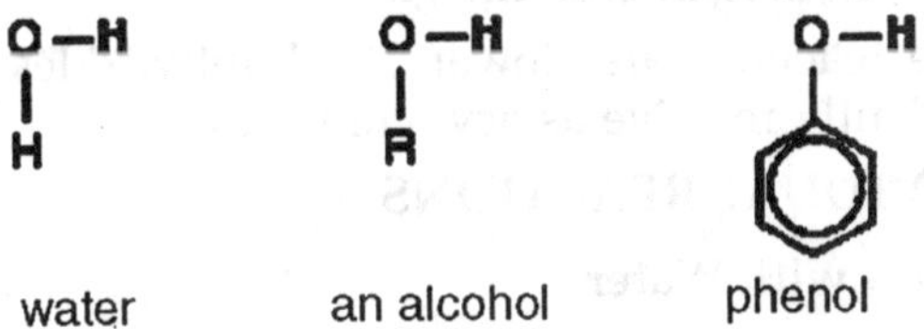

water an alcohol phenol

Comparing the reactions of acyl chlorides and acid anhydrides with these compounds

Because the formula is much easier, it helps to start with the acyl chlorides.

The Reactions with acyl Chlorides

We'll take ethanoyl chloride as typical of the acyl chlorides.

Taking a general case of a reaction between ethanoyl chloride and a compound X-O-H (where X is hydrogen, or an alkyl group or a benzene ring):

$$CH_3\text{-}C(=O)\text{-}Cl \;+\; X\text{-}O\text{-}H \longrightarrow CH_3\text{-}C(=O)\text{-}O\text{-}X \;+\; HCl$$

So in each case, hydrogen chloride gas is produced - the hydrogen coming from the -OH group, and the chlorine from the ethanoyl chloride. Everything left over just gets joined together.

The same reactions with acid anhydrides

$$CH_3\text{-}C(=O)\text{-}O\text{-}C(=O)\text{-}CH_3 \;+\; X\text{-}O\text{-}H \longrightarrow CH_3\text{-}C(=O)\text{-}O\text{-}X \;+\; CH_3COOH$$

If you compare this with the acyl chloride equation, you can see that the only difference is that ethanoic acid is produced as the second product of the reaction rather than hydrogen chloride.

These reactions are just the same as the corresponding acyl chloride reactions except:

- Ethanoic acid is formed as the second product rather than hydrogen chloride gas.
- The reactions are slower. Acid anhydrides aren't so violently reactive as acyl chlorides.

THE INDIVIDUAL REACTIONS

The Reaction with Water

Modifying the general equation we've just looked at, you will see that you just get two molecules of ethanoic acid produced.

$$CH_3\text{-}C(=O)\text{-}O\text{-}C(=O)\text{-}CH_3 \;+\; H\text{-}O\text{-}H \longrightarrow CH_3\text{-}C(=O)\text{-}O\text{-}H \;+\; CH_3COOH$$

This is more usually (and more easily!) written as:

$$(CH_3CO)_2O + H_2O \longrightarrow 2CH_3COOH$$

The reaction happens slowly at room temperature (faster on gentle warming) without a great deal exciting to observe - unlike in the acyl chloride case where hydrogen chloride fumes are produced. You mix two colourless liquids and get another colourless liquid.

The equivalent acyl chloride reaction is:

The Reaction with Alcohols

We'll start by taking the general case of any alcohol reacting with ethanoic anhydride. The equation would be:

$$CH_3C(=O)-O-C(=O)CH_3 + R-O-H \longrightarrow CH_3C(=O)-O-R + CH_3COOH$$

or, more simply:

$$(CH_3CO)_2O + ROH \longrightarrow \underset{\text{an ester}}{CH_3COOR} + \underset{\text{ethanoic acid}}{CH_3COOH}$$

The product this time (apart from the ethanoic acid always produced) is an ester. For example, with ethanol you would get the ester ethyl ethanoate:

$$(CH_3CO)_2O + CH_3CH_2OH \longrightarrow \underset{\text{ethylethanoate}}{CH_3COOCH_2CH_3} + CH_3COOH$$

This reaction also needs gentle heating for it to happen at a reasonable rate, and again there isn't anything visually dramatic.

The equivalent acyl chloride reaction is:

$$CH_3COCl + CH_3CH_2OH \longrightarrow CH_3COOCH_2CH_3 + HCl$$

The Reaction with Phenols

The Reaction with Phenol itself

Phenols have an -OH group attached directly to a benzene ring. In the substance normally called "phenol", there isn't anything else attached to the ring as well. We'll look at that first.

The reaction between phenol and ethanoic anhydride isn't particularly important, but you would get an ester just as you do with an alcohol.

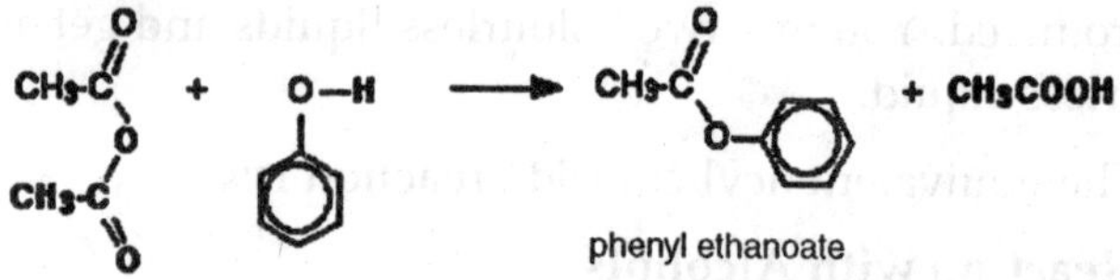

phenyl ethanoate

Or, more simply:

$$(CH_3CO)_2O + C_6H_5OH \longrightarrow CH_3COOC_6H_5 + CH_3COOH$$

Especially if you write the equation in this second way, it is obvious that you have just produced another ester - in this case, called phenyl ethanoate.

The equivalent acyl chloride reaction is:

$$CH_3COCl + C_6H_5OH \longrightarrow CH_3COOC_6H_5 + HCl$$

You may come across the structure of the ester drawn in a variety of other ways which make it look much more as if it was a derivative of phenol.

For example:

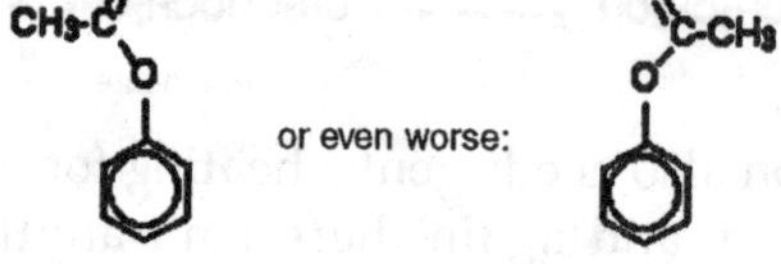

or even worse:

Looking at it this way, notice that the hydrogen of the phenol -OH group has been replaced by an acyl group - an alkyl group attached to a carbon-oxygen double bond.

You can say that the phenol has been acylated or has undergone acylation.

Because of the nature of this particular acyl group, it is also described as ethanoylation. The hydrogen is being replaced by an ethanoyl group, CH_3CO-.

Using a similar reaction to make aspirin

The reaction with phenol itself isn't very important, but you can make aspirin by a very similar reaction.

The molecule below is 2-hydroxybenzoic acid (also known as 2-hydroxybenzenecarboxylic acid). The old name for this is salicylic acid.

You might find it written in either of these two ways. They are the same structure with the molecule just flipped over in space.

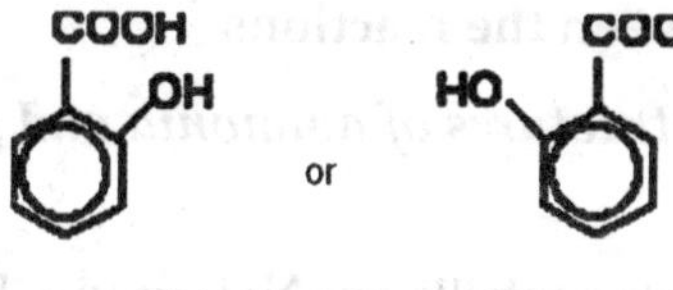

COOH OH or HO COOH

2-hydroxybanzoic acid

You might also find it with the -OH group at the top and the -COOH group next door and either to the left or right of it. Life can get very confusing!

When this reacts with ethanoic anhydride, it is ethanoylated (or acylated, if you want to use the more general term) to give:

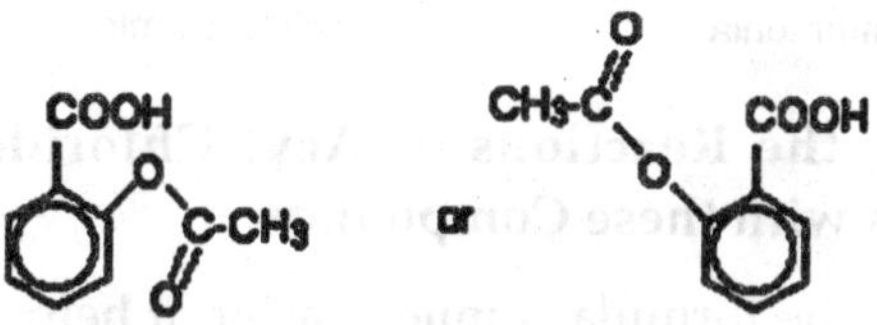

COOH O C-CH3 O or O CH3-C O COOH

You might find all sorts of other variants on drawing this as well.

This molecule is aspirin.

Although this reaction can also be done with ethanoyl chloride, aspirin is manufactured by reacting 2-hydroxybenzoic acid with ethanoic anhydride at 90°C.

The reasons for using ethanoic anhydride rather than ethanoyl chloride include:

- Ethanoic anhydride is cheaper than ethanoyl chloride.
- Ethanoic anhydride is safer to use than ethanoyl chloride. It is less corrosive and not so readily hydrolyzed (its reaction with water is slower).
- Ethanoic anhydride doesn't produce dangerous (corrosive and poisonous) fumes of hydrogen chloride.

REACTIONS OF ACID ANHYDRIDES WITH AMMONIA AND PRIMARY AMINES

Similarities between the reactions

Comparing the structures of ammonia and primary amines

Each substance contains an -NH_2 group. In ammonia, this is attached to a hydrogen atom. In a primary amine, it is attached to an alkyl group (shown by "R" in the diagram below) or a benzene ring.

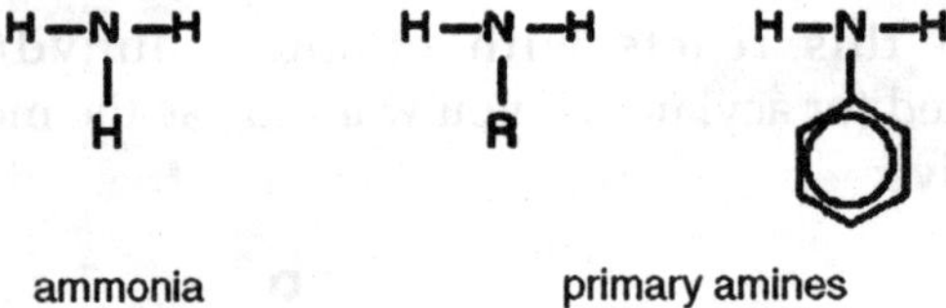

Comparing the Reactions of Acyl Chlorides and acid Anhydrides with these Compounds

Because the formula is much easier, it helps to start with the acyl chlorides.

The Reactions with Acyl Chlorides

We'll take ethanoyl chloride as typical of the acyl chlorides.

Taking a general case of a reaction between ethanoyl chloride and a compound XNH_2 (where X is hydrogen, or an alkyl group, or a benzene ring). The reaction happens in two stages:

First:

$$CH_3-C(=O)-Cl \; + \; H-N(X)-H \longrightarrow CH_3-C(=O)-N(H)-X \; + \; HCl$$

So in each case, you initially get hydrogen chloride gas - the hydrogen coming from the -NH_2 group, and the chlorine from the ethanoyl chloride. Everything left over just gets joined together.

But ammonia and amines are basic, and react with the hydrogen chloride to produce a salt. So the second stage of the reaction is:

$$XNH_2 + HCl \longrightarrow XNH_3^+ \; Cl^-$$

The same reactions with acid anhydrides

Again, the reaction happens in two stages. In the first:

$$CH_3-C(=O)-O-C(=O)-CH_3 \; + \; H-N(X)-H \longrightarrow CH_3-C(=O)-N(H)-X \; + \; CH_3COOH$$

If you compare this with the acyl chloride equation, you can see that the only difference is that ethanoic acid is produced as the second product of the reaction rather than hydrogen chloride.

Then the ethanoic acid reacts with excess ammonia or amine to give a salt - this time an ethanoate.

$$CH_3COOH + XNH_2 \longrightarrow CH_3COO^- + NH_3X$$

This looks more difficult than the acyl chloride case because of the way the salt is written. You get an ethanoate ion and a positive ion with this structure:

$$H-\overset{+}{N}(H)(H)-X$$

This is easier to understand with real compounds - as you will see below.

In summary:

These reactions are just the same as the corresponding acyl chloride reactions except:

- Initially, ethanoic acid is formed as the second product rather than hydrogen chloride gas.
- The second stage of the reaction involves the formation of an ethanoate rather than a chloride.
- The reactions are slower. Acid anhydrides aren't so violently reactive as acyl chlorides, and the reactions normally need heating.

THE INDIVIDUAL REACTIONS

The Reaction with Ammonia

In this case, the "X" in the equations above is a hydrogen atom. So in the first instance you get ethanoic acid and an organic compound called an amide.

Amides contain the group -$CONH_2$. In the reaction between ethanoic anhydride and ammonia, the amide formed is called ethanamide.

$$CH_3\text{-}C(=O)\text{-}O\text{-}C(=O)\text{-}CH_3 + H\text{-}N(H)\text{-}H \longrightarrow CH_3\text{-}C(=O)\text{-}N(H)\text{-}H + CH_3COOH$$

This is more usually written as:

$$(CH_3CO)_2O + NH_3 \longrightarrow \underset{\text{ethanamide}}{CH_3CONH_2} + CH_3CCCH$$

The ethanoic acid produced reacts with excess ammonia to give ammonium ethanoate.

$$CH_3COOH + NH_3 \longrightarrow \underset{\text{ammonium ethanoate}}{CH_3COO^- + NH_4}$$

and you can combine all this together to give one overall equation:

$$(CH_3CO)_2O + 2NH_3 \longrightarrow \underset{\text{ethanamide}}{CH_3CONH_2} + \underset{\text{ammonium ethanoate}}{CH_3COO^- \, {}^+NH_4}$$

You need to follow this through really carefully, because the two products of the reaction overall can look confusingly similar.

The corresponding reaction with an acyl chloride is:

$$CH_3COCl + 2NH_3 \longrightarrow CH_3CONH_2 + NH_4^+Cl^-$$

The Reaction with Primary Amines

The Reaction with Methylamine

We'll take methylamine as typical of primary amines where the $-NH_2$ is attached to an alkyl group.

The initial equation would be:

$$CH_3-C(=O)-O-C(=O)-CH_3 + H-N(CH_3)-H \longrightarrow CH_3-C(=O)-N(H)-CH_3 + CH_3COOH$$

The first product this time is called an N-substituted amide.

If you compare the structure with the amide produced in the reaction with ammonia, the only difference is that one of the hydrogens on the nitrogen has been substituted for a methyl group.

This particular compound is N-methylethanamide. The "N" simply shows that the substitution is on the nitrogen atom, and not elsewhere in the molecule.

The equation would normally be written:

$$(CH_3CO)_2O + CH_3NH_2 \longrightarrow \underset{\text{N-methylethanamide}}{CH_3CONHCH_3} + CH_3COOH$$

You can think of primary amines as just being modified ammonia. If ammonia is basic and forms a salt with the ethanoic acid, excess methylamine will do exactly the same thing.

$$CH_3COOH + CH_3NH_2 \longrightarrow CH_3COO^- \; {}^+NH_3CH_3$$

methylammonium ethanoate

The salt is called methylammonium ethanoate. It is just like ammonium ethanoate, except that one of the hydrogens has been replaced by a methyl group.

You would usually combine these equations into one overall equation for the reaction:

$$(CH_3CO)_2O + 2CH_3NH_2 \longrightarrow CH_3CONHCH_3 + CH_3COO- \; {}^+NH_3CH_3$$

The corresponding reaction with an acyl chloride is:

$$CH_3COCl + 2CH_3NH_2 \longrightarrow CH_3CONHCH_3{}^+ \; CH_3NH_3{+}Cl^-$$

The Reaction with Phenylamine (aniline)

Phenylamine is the simplest primary amine where the -NH_2 group is attached directly to a benzene ring. Its old name is aniline.

In phenylamine, there isn't anything else attached to the ring as well. You can write the formula of phenylamine as $C_6H_5NH_2$.

There is no essential difference between this reaction and the reaction with methylamine, but the structure of the N-substituted amide formed is different.

The overall equation for the reaction is:

$$(CH_3CO)_2O + 2C_6H_5NH_2 \longrightarrow CH_3CONHC_6H_5 + CH_3COO- + NH_3C_6H_5$$

The products are N-phenylethanamide and phenylammonium ethanoate.

This reaction can sometimes look confusing if the phenylamine is drawn showing the benzene ring, and especially if the reaction is looked at from the point of view of the phenylamine.

For example, the product molecule might be drawn looking like this:

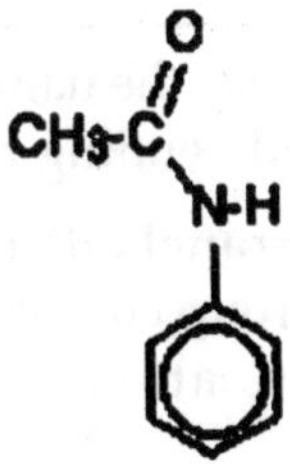

If you stop and think about it, this is obviously the same molecule as in the equation above, but it stresses the phenylamine part of it much more.

Looking at it this way, notice that one of the hydrogens of the $-NH_2$ group has been replaced by an acyl group - an alkyl group attached to a carbon-oxygen double bond.

You can say that the phenylamine has been acylated or has undergone acylation.

Because of the nature of this particular acyl group, it is also described as ethanoylation. The hydrogen is being replaced by an ethanoyl group, CH_3CO-.

AMIDES

Introduction

Amides are derived from carboxylic acids. A carboxylic acid contains the -COOH group, and in an amide the -OH part of that group is replaced by an $-NH_2$ group.

So amides contain the $-CONH_2$ group.

Some Simple Amides

The most commonly discussed amide is ethanamide, CH_3CONH_2 (old name: acetamide).

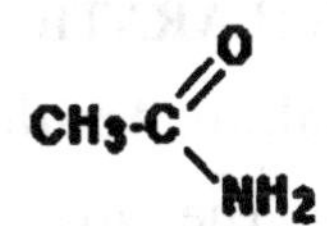

ethanamide

The three simplest amides are:

$HCONH_2$	Methanamide
CH_3CONH_2	Ethanamide
$CH_3CH_2CONH_2$	propanamide

Notice that in each case, the name is derived from the acid by replacing the "oic acid" ending by "amide".

If the chain was branched, the carbon in the -$CONH_2$ group counts as the number 1 carbon atom. For example:

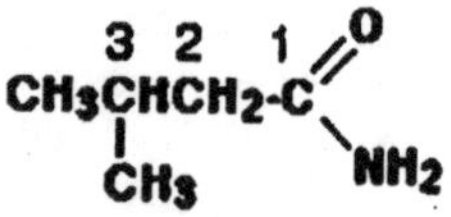

$$\overset{3}{CH_3}\underset{CH_3}{\overset{2}{CH}}\overset{1}{CH_2}\text{-}C(=O)NH_2$$

3-methylbutanamide

PHYSICAL PROPERTIES

Melting Points

Methanamide is a liquid at room temperature (melting point: 3°C), but the other amides are solid.

For example, ethanamide forms colourless deliquescent crystals with a melting point of 82°C. A deliquescent substance is one which picks up water from the atmosphere and dissolves in it. Ethanamide crystals nearly always look wet. The melting points of the amides are high for the size of the molecules because they can form hydrogen bonds.

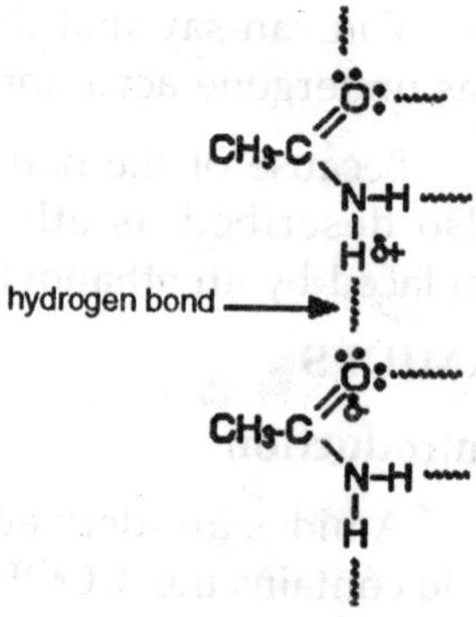

hydrogen bond

The hydrogen atoms in the -NH_2 group are sufficiently positive to form a hydrogen bond with a lone pair on the oxygen atom of another molecule.

PREPARATION

Making Amides from Carboxylic Acids

The carboxylic acid is first converted into an ammonium salt which then produces an amide on heating.

The ammonium salt is formed by adding solid ammonium carbonate to an excess of the acid.

For example, ammonium ethanoate is made by adding ammonium carbonate to an excess of ethanoic acid.

$$2CH_3COOH + (NH_4)_2CO_3 \longrightarrow 2CH_3COONH_4 + H_2O + CO_2$$

When the reaction is complete, the mixture is heated and the ammonium salt dehydrates producing ethanamide.

$$CH_3COONH_4 \longrightarrow CH_3CONH_2 + H_2O$$

The excess of ethanoic acid is there to prevent dissociation of the ammonium salt before it dehydrates.

Ammonium salts tend to split into ammonia and the parent acid on heating, recombining on cooling. If dissociation happened in this case, the ammonia would escape from the reaction mixture and be lost. You couldn't get any recombination.

The dissociation is reversible:

$$CH_3COONH_{4(s)} \rightleftharpoons CH_3COOH_{(l)} + NH_{3(g)}$$

The presence of the excess ethanoic acid helps to prevent this from happening by moving the position of equilibrium to the left.

Some Brief Practical Details

The ammonium carbonate is added slowly to concentrated ethanoic acid and the reaction is left until all production of carbon dioxide stops.

It is then heated under reflux for half an hour for the dehydration to take place.

The mixture is distilled at about 170°C to remove excess ethanoic acid and water - leaving almost pure ethanamide in the flask.

Making Amides from Acyl Chlorides

Acyl chlorides (also known as acid chlorides) have the general formula RCOCl. The chlorine atom is very easily replaced by other things. For example, it is easily replaced by an -NH_2 group to make an amide.

To make ethanamide from ethanoyl chloride, you normally add the ethanoyl chloride to a concentrated solution of ammonia in water. There is a very violent reaction producing lots of white smoke - a mixture of solid ammonium chloride and ethanamide. Some of the mixture remains dissolved in water as a colourless solution.

You can think of the reaction as happening in two stages.

In the first stage, the ammonia reacts with the ethanoyl chloride to give ethanamide and hydrogen chloride gas.

$$CH_3COCl + NH_3 \longrightarrow CH_3CONH_2 + HCl$$

Then the hydrogen chloride produced reacts with excess ammonia to give ammonium chloride.

$$NH_3 + HCl \longrightarrow NH_4Cl$$

and you can combine all this together to give one overall equation:

$$CH_3COCl + 2NH_3 \longrightarrow CH_3CONH_2 + NH_4Cl$$

Making Amides from Acid Anhydrides

An acid anhydride is what you get if you remove a molecule of water from two carboxylic acid -COOH groups.

For example, if you took two ethanoic acid molecules and removed a molecule of water between them you would get the acid anhydride, ethanoic anhydride (old name: acetic anhydride).

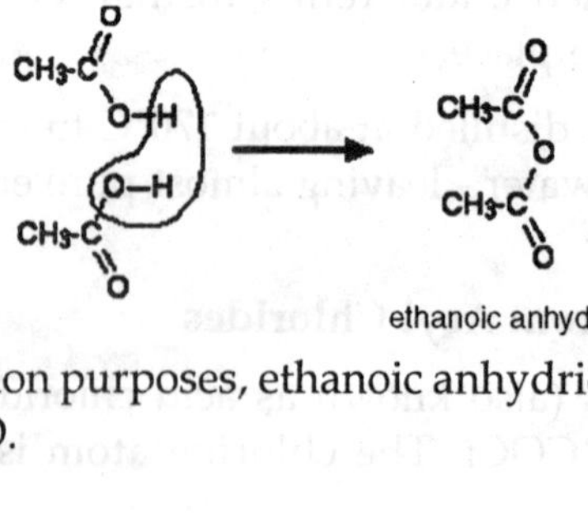

ethanoic anhydride

For equation purposes, ethanoic anhydride is often written as $(CH_3CO)_2O$.

The reactions of acid anhydrides are rather like those of acyl chlorides except that during their reactions, a molecule of carboxylic acid is produced rather than the HCl formed when an acyl chloride reacts.

If ethanoic anhydride is added to concentrated ammonia solution, ethanamide is formed together with ammonium ethanoate. Again, the reaction happens in two stages.

In the first stage, ethanamide is formed together with ethanoic acid.

$$(CH_3CO)_2O + NH_3 \longrightarrow CH_3CONH_2 + CH_3COOH$$

Then the ethanoic acid produced reacts with excess ammonia to give ammonium ethanoate.

$$CH_3COOH + NH_3 \longrightarrow CH_3COONH_4$$

and you can combine all this together to give one overall equation:

$$(CH_3CO)_2O + 2NH_3 \longrightarrow CH_3CONH_2 + CH_3COONH_4$$

You need to follow this through really carefully, because the two products of the reaction overall can look confusingly similar.

REACTIONS

Hydrolysis

Technically, hydrolysis is a reaction with water. That is exactly what happens when amides are hydrolyzed in the presence of dilute acids such as dilute hydrochloric acid. The acid acts as a catalyst for the reaction between the amide and water.

The alkaline hydrolysis of amides actually involves reaction with hydroxide ions, but the result is similar enough that it is still classed as hydrolysis.

Hydrolysis under acidic Conditions

Taking ethanamide as a typical amide:

If ethanamide is heated with a dilute acid (such as dilute hydrochloric acid), ethanoic acid is formed together with ammonium ions. So, if you were using hydrochloric acid, the final solution would contain ammonium chloride and ethanoic acid.

$$CH_3CONH_2 + H_2O + HCl \longrightarrow CH_3COOH + NH_4^+ Cl^-$$

Hydrolysis under Alkaline Conditions

Again, taking ethanamide as a typical amide:

If ethanamide is heated with sodium hydroxide solution, ammonia gas is given off and you are left with a solution containing sodium ethanoate.

$$CH_3CONH_2 + NaOH \longrightarrow CH_3COONa + NH_3$$

Using Alkaline Hydrolysis to Test for an Amide

If you add sodium hydroxide solution to an unknown organic compound, and it gives off ammonia on heating (but not immediately in the cold), then it is an amide.

You can recognize the ammonia by smell and because it turns red litmus paper blue.

The possible confusion using this test is with ammonium salts. Ammonium salts also produce ammonia with sodium hydroxide solution, but in this case there is always enough ammonia produced in the cold for the smell to be immediately obvious.

As you can see, there is the potential for lots of hydrogen bonds to be formed. Each molecule has two slightly positive hydrogen atoms and two lone pairs on the oxygen atom.

These hydrogen bonds need a reasonable amount of energy to break, and so the melting points of the amides are quite high.

Solubility in Water

The small amides are soluble in water because they have the ability to hydrogen bond with the water molecules.

It needs energy to break the hydrogen bonds between amide molecules and between water molecules before they can mix - but enough energy is released again when the new hydrogen bonds are set up to allow this to happen.*

OTHER REACTIONS OF AMIDES

The Lack of base Character in Amides

Unusually for compounds containing the $-NH_2$ group, amides are neutral. This section explains why $-NH_2$ groups are usually basic and why amides are different.

The usual basic Character of the $-NH_2$ group

Simple compounds containing an $-NH_2$ group such as ammonia, NH_3, or a primary amine like methylamine, CH_3NH_2, are weak bases. A primary amine is a compound where the $-NH_2$ group is attached to a hydrocarbon group.

The active lone pair of electrons on the nitrogen atom in ammonia can combine with a hydrogen ion (a proton) from some other source - in other words it acts as a base.

With a compound like methylamine, all that has happened is that one of the hydrogen atoms attached to the nitrogen has been replaced by a methyl group. It doesn't make a huge amount of difference to the lone pair and so ammonia and methylamine behave similarly.

For example, if you dissolve these compounds in water, the nitrogen lone pair takes a hydrogen ion from a water molecule - and equilibria like these are set up:

$$NH_{3(aq)} + H_2O_{(l)} \rightleftharpoons NH_4^+{}_{(aq)} + OH^-{}_{(aq)}$$

$$CH_3NH_{2(aq)} + H_2O_{(l)} \rightleftharpoons CH_3NH_3^+{}_{(aq)} + OH^-{}_{(aq)}$$

Notice that the reactions are reversible. In both cases the positions of equilibrium lie well to the left. These compounds are weak bases because they don't hang on to the incoming hydrogen ion very well.

Both ammonia and the amines are alkaline in solution because of the presence of the hydroxide ions, and both of them turn red litmus blue.

Why doesn't Something Similar Happen with Amides?

Amides are neutral to litmus and have virtually no basic character at all - despite having the $-NH_2$ group. Their tendency to attract hydrogen ions is so slight that it can be ignored for most purposes.

We need to look at the bonding in the $-CONH_2$ group.

Like any other double bond, a carbon-oxygen double bond is made up of two different parts. One electron pair is found on the line between the two nuclei - this is known as a sigma bond. The other electron pair is found above and below the plane of the molecule in a pi bond.

A pi bond is made by sideways overlap between p orbitals on the carbon and the oxygen.

In an amide, the lone pair on the nitrogen atom ends up almost parallel to these p orbitals, and overlaps with them as they form the pi bond.

The result of this is that the nitrogen lone pair becomes delocalized - in other words it is no longer found located on the nitrogen atom, but the electrons from it are spread out over the whole of that part of the molecule.

This has two effects which prevent the lone pair accepting hydrogen ions and acting as a base:

- Because the lone pair is no longer located on a single atom as an intensely negative region of space, it isn't anything like as attractive for a nearby hydrogen ion.
- Delocalization makes molecules more stable. For the nitrogen to reclaim its lone pair and join to a hydrogen ion, the delocalization would have to be broken, and that will cost energy.

Dehydration

Amides are dehydrated by heating a solid mixture of the amide and phosphorus(V) oxide, P_4O_{10}.

Water is removed from the amide group to leave a nitrile group, -CN. The liquid nitrile is collected by simple distillation.

For example, with ethanamide, you will get ethanenitrile.

$$CH_3CONH_2 \xrightarrow[-H_2O]{P_4O_{10}} CH_3CN$$

The Hofmann Degradation

The Hofmann degradation is a reaction between an amide and a mixture of bromine and sodium hydroxide solution. Heat is needed.

The net effect of the reaction is a loss of the -CO- part of the amide group. You get a primary amine with one less carbon atom than the original amide had.

The general case would be (as a flow scheme):

$$RCONH_2 \xrightarrow{Br_2\ /\ NaOH} RNH_2$$

If you started with ethanamide, you would get methylamine. The full equation for the reaction is:

$$CH_3CONH_2 + Br_2 + 4NaOH \longrightarrow CH_3NH_2 + Na_2CO_3 + 2NaBr + 2H_2O$$

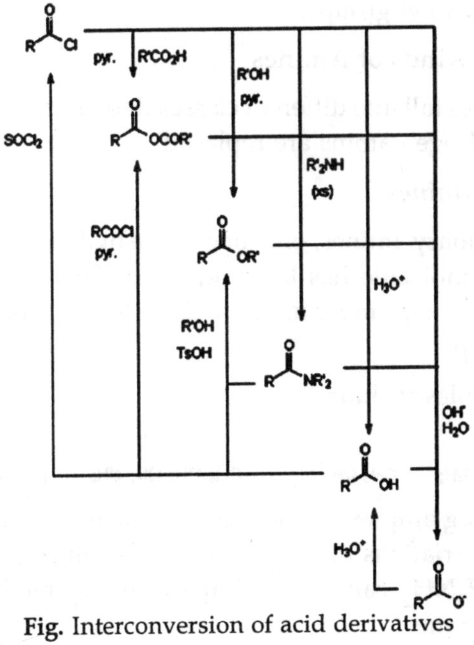

Fig. Interconversion of acid derivatives

Chapter 14

Amines

INTRODUCTION

The easiest way to think of amines is as near relatives of ammonia, NH_3.

In amines, the hydrogen atoms in the ammonia have been replaced one at a time by hydrocarbon groups. On this page, we are only looking at cases where the hydrocarbon groups are simple alkyl groups.

Different Kinds of Amines

Amines fall into different classes depending on how many of the hydrogen atoms are replaced.

Primary Amines

In primary amines, only one of the hydrogen atoms in the ammonia molecule has been replaced. That means that the formula of the primary amine will be RNH_2 where "R" is an alkyl group.

Examples include:

$CH_3\text{-}NH_2$ $\quad$ $CH_3\text{-}CH_2\text{-}NH_2$ $\quad$ $CH_3\text{-}CH_2\text{-}CH_2\text{-}NH_2$ $\quad$ $CH_3\text{-}\underset{}{CH}(NH_2)\text{-}CH_3$

Naming amines can be quite confusing because there are so many variations on the names. For example, the simplest amine, CH_3NH_2, can be called methylamine, methanamine or aminomethane.

$$CH_3\text{-}NH_2 = \begin{array}{l}\text{methylamine}\\ \text{methanamine}\\ \text{aminomethane}\end{array}$$

The commonest name at this level is methylamine and, similarly, the second compound drawn above is usually called ethylamine.

Where there might be confusion about where the $-NH_2$ group is attached to a chain, the simplest way of naming the compound is to use the "amino" form.

For example:

$CH_3\text{-}CH_2\text{-}CH_2\text{-}NH_2$ = 1-aminopropane

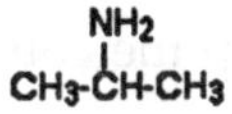

$CH_3\text{-}CH(NH_2)\text{-}CH_3$ = 2-aminopropane

Secondary Amines

In a secondary amine, two of the hydrogens in an ammonia molecule have been replaced by hydrocarbon groups. At this level, you are only likely to come across simple ones where both of the hydrocarbon groups are alkyl groups and both are the same.

For example:

$CH_3\text{-}NH\text{-}CH_3$ dimethylamine

$CH_3\text{-}CH_2\text{-}NH\text{-}CH_2\text{-}CH_3$ diethylamine

There are other variants on the names, but this is the commonest and simplest way of naming these small secondary amines.

Tertiary Amines

In a tertiary amine, all of the hydrogens in an ammonia molecule have been replaced by hydrocarbon groups. Again, you are only likely to come across simple ones where all three of the hydrocarbon groups are alkyl groups and all three are the same.

The naming is similar to secondary amines.

For example:

$$CH_3-\underset{}{\overset{CH_3}{\overset{|}{N}}}-CH_3$$

trimethylamine

PREPARATION

From Halogenoalkanes

The halogenoalkane is heated with a concentrated solution of ammonia in ethanol. The reaction is carried out in a sealed tube. You couldn't heat this mixture under reflux, because the ammonia would simply escape up the condenser as a gas.

We'll talk about the reaction using 1-bromoethane as a typical halogenoalkane.

You get a mixture of amines formed together with their salts. The reactions happen one after another.

Making a Primary Amine

The reaction happens in two stages. In the first stage, a salt is formed - in this case, ethylammonium bromide. This is just like ammonium bromide, except that one of the hydrogens in the ammonium ion is replaced by an ethyl group.

$$CH_3CH_2Br + NH_3 \longrightarrow CH_3CH_2NH_3 + Br^-$$

There is then the possibility of a reversible reaction between this salt and excess ammonia in the mixture.

$$CH_3CH_2NH_3 + Br^- + NH_3 \rightleftharpoons CH_3CH_2NH_2 + NH_4 + Br^-$$

The ammonia removes a hydrogen ion from the ethylammonium ion to leave a primary amine - ethylamine.

The more ammonia there is in the mixture, the more the forward reaction is favoured.

Making a Secondary Amine

The reaction doesn't stop at a primary amine. The ethylamine also reacts with bromoethane - in the same two stages as before.

In the first stage, you get a salt formed - this time, diethylammonium bromide. Think of this as ammonium bromide with two hydrogens replaced by ethyl groups.

$$CH_3CH_2Br + CH_3CH_2NH_2 \longrightarrow (CH_3CH_2)_2NH_2^+ \; Br^-$$

There is again the possibility of a reversible reaction between this salt and excess ammonia in the mixture.

$$(CH_3CH_2)_2NH_2^+ \; Br^- + NH_3 \rightleftharpoons (CH_3CH_2)_2NH + NH_4^+ Br^-$$

The ammonia removes a hydrogen ion from the diethylammonium ion to leave a secondary amine - diethylamine. A secondary amine is one which has two alkyl groups attached to the nitrogen.

Making a Tertiary Amine

The diethylamine also reacts with bromoethane - in the same two stages as before.

In the first stage, you get triethylammonium bromide.

$$(CH_3CH_2)_3NH^+ \; Br^- + NH_3 \rightleftharpoons (CH_3CH_2)_3N + NH_4^+ Br^-$$

There is again the possibility of a reversible reaction between this salt and excess ammonia in the mixture.

$$(CH_3CH_2)_3NH^+ \; Br^- + NH_3 \rightleftharpoons (CH_3CH_2)_3N + NH_4^+ Br^-$$

The ammonia removes a hydrogen ion from the triethylammonium ion to leave a tertiary amine - triethylamine. A tertiary amine is one which has three alkyl groups attached to the nitrogen.

Making a Quaternary Ammonium Salt

The final stage! The triethylamine reacts with bromoethane to give tetraethylammonium bromide - a quaternary ammonium salt (one in which all four hydrogens have been replaced by alkyl groups).

$$CH_3CH_2Br + (CH_3CH_2)_3N \longrightarrow (CH_3CH_2)_4N^+ \; Br^-$$

This time there isn't any hydrogen left on the nitrogen to be removed. The reaction stops here.

What do you Actually get if you React Bromoethane with Ammonia?

Whatever you do, you get a mixture of all of the products (including the various amines and their salts).

To get mainly the quaternary ammonium salt, you can use a large excess of bromoethane. If you look at the reactions going on, each one needs additional bromoethane. If you provide enough, then the chances are that the reaction will go to completion, given enough time.

On the other hand, if you use a very large excess of ammonia, the chances are always greatest that a bromoethane molecule will hit an ammonia molecule rather than one of the amines being formed. That will help to prevent the formation of secondary (etc) amines - although it won't stop it entirely.

FROM NITRILES

Making Primary Amines from Nitriles

Nitriles are compounds containing the -CN group, and can be reduced in various ways. Two possible methods are described here.

Reducing Nitriles using $LiAlH_4$

One possible reducing agent is lithium tetrahydridoaluminate(III) - often just called lithium tetrahydridoaluminate or lithium aluminium hydride.

The nitrile reacts with the lithium tetrahydridoaluminate in solution in ethoxyethane (diethyl ether, or just "ether") followed by treatment of the product of that reaction with a dilute acid.

Overall, the carbon-nitrogen triple bond is reduced to give a primary amine.

For example, with ethanenitrile you get ethylamine:

$$CH_3CN + 4(H) \longrightarrow CH_3CH_2NH_2$$

Notice that this is a simplified equation. [H] means "hydrogen from a reducing agent".

The Reduction of Nitriles using Hydrogen and a Metal Catalyst

The carbon-nitrogen triple bond in a nitrile can also be reduced by reaction with hydrogen gas in the presence of a variety of metal catalysts.

Commonly quoted catalysts are palladium, platinum or nickel.

The reaction will take place at a raised temperature and pressure. It is impossible to give exact details because it will vary from catalyst to catalyst.

For example, ethanenitrile can be reduced to ethylamine by reaction with hydrogen in the presence of a palladium catalyst.

$$CH_3CN + 2H_2 \xrightarrow{Pd} CH_3CH_2NH_2$$

Physical Properties of Amines

Boiling points

The table shows the boiling points of some simple amines.

type	*Formula*	*Boiling point (°C)*
primary	CH_3NH_2	-6.3
primary	$CH_3CH_2NH_2$	16.6
primary	$CH_3CH_2CH_2NH_2$	48.6
secondary	$(CH_3)_2NH$	7.4
tertiary	$(CH_3)_3N$	3.5

We will need to look at this with some care to sort out the patterns and reasons. Concentrate first on the primary amines.

Primary Amines

It is useful to compare the boiling point of methylamine, CH_3NH_2, with that of ethane, CH_3CH_3.

Both molecules contain the same number of electrons and have, as near as makes no difference, the same shape. However, the boiling point of methylamine is -6.3°C, whereas ethane's boiling point is much lower at -88.6°C.

The reason for the higher boiling points of the primary amines is that they can form hydrogen bonds with each other as well as Van der Waals dispersion forces and dipole-dipole interactions.

Hydrogen bonds can form between the lone pair on the very electronegative nitrogen atom and the slightly positive hydrogen atom in another molecule.

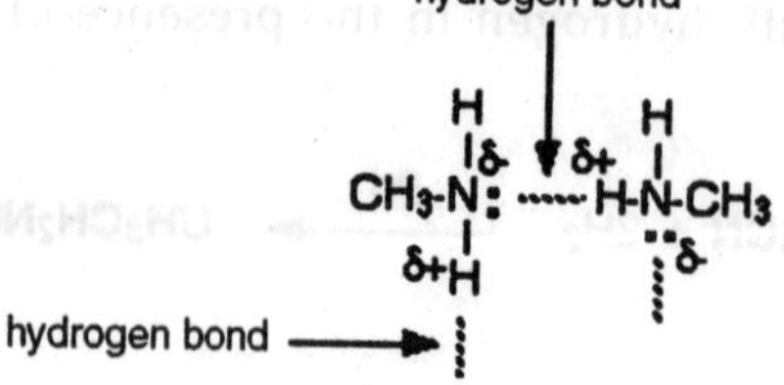

The hydrogen bonding isn't as efficient as it is in, say, water, because there is a shortage of lone pairs. Some slightly positive hydrogen atoms won't be able to find a lone pair to hydrogen bond with. There are twice as many suitable hydrogens are there are lone pairs.

The boiling points of the primary amines increase as you increase chain length because of the greater amount of Van der Waals dispersion forces between the bigger molecules.

Secondary Amines

For a fair comparison you would have to compare the boiling point of dimethylamine with that of ethylamine. They are isomers of each other - each contains exactly the same number of the same atoms.

The boiling point of the secondary amine is a little lower than the corresponding primary amine with the same number of carbon atoms.

Secondary amines still form hydrogen bonds, but having the nitrogen atom in the middle of the chain rather than at the end makes the permanent dipole on the molecule slightly less.

The lower boiling point is due to the lower dipole-dipole attractions in the dimethylamine compared with ethylamine.

Tertiary Amines

This time to make a fair comparison you would have to compare trimethylamine with its isomer 1-aminopropane.

If you look back at the table further up the page, you will see that the trimethylamine has a much lower boiling point (3.5°C) than 1-aminopropane (48.6°C).

In a tertiary amine there aren't any hydrogen atoms attached directly to the nitrogen. That means that hydrogen bonding between tertiary amine molecules is impossible. That's why the boiling point is much lower.

Solubility in Water

The small amines of all types are very soluble in water. In fact, the ones that would normally be found as gases at room temperature are normally sold as solutions in water - in much the same way that ammonia is usually supplied as ammonia solution.

All of the amines can form hydrogen bonds with water - even the tertiary ones.

Although the tertiary amines don't have a hydrogen atom attached to the nitrogen and so can't form hydrogen bonds with themselves, they can form hydrogen bonds with water molecules just using the lone pair on the nitrogen.

hydrogen bond

CH_3 | O–H

CH_3–$\ddot{N}^{\delta-}$ ······ $H^{\delta+}$

CH_3

Solubility falls off as the hydrocarbon chains get longer - noticeably so after about 6 carbons. The hydrocarbon chains have to force their way between water molecules, breaking hydrogen bonds between water molecules.

However, they don't replace them by anything as strong, and so the process of forming a solution becomes less and less energetically feasible as chain length grows.

Smell

The very small amines like methylamine and ethylamine smell very similar to ammonia - although if you compared them side by side, the amine smells are slightly more complex.

As the amines get bigger, they tend to smell more "fishy", or they smell of decay.

If you are familiar with the smell of hawthorn blossom (and similarly smelling things like cotoneaster blossom), this is the smell of trimethylamine - a sweet and rather sickly smell like the early stages of decaying flesh.

BASIC PROPERTIES OF AMINES

We are going to have to use two different definitions of the term "base" in this page.

A base is

- a substance which combines with hydrogen ions. This is the Bronsted-Lowry theory.
- an electron pair donor. This is the Lewis theory.

The easiest way of looking at the basic properties of amines is to think of an amine as a modified ammonia molecule. In an amine, one or more of the hydrogen atoms in ammonia has been replaced by a hydrocarbon group.

Replacing the hydrogens still leaves the lone pair on the nitrogen unchanged - and it is the lone pair on the nitrogen that gives ammonia its basic properties. Amines will therefore behave much the same as ammonia in all cases where the lone pair is involved.

REACTIONS OF AMINES WITH ACIDS

These are most easily considered using the Bronsted-Lowry theory of acids and bases - the base is a hydrogen ion acceptor. We'll do a straight comparison between amines and the familiar ammonia reactions.

A Reminder about the Ammonia Reactions

Ammonia reacts with acids to produce ammonium ions. The ammonia molecule picks up a hydrogen ion from the acid and attaches it to the lone pair on the nitrogen.

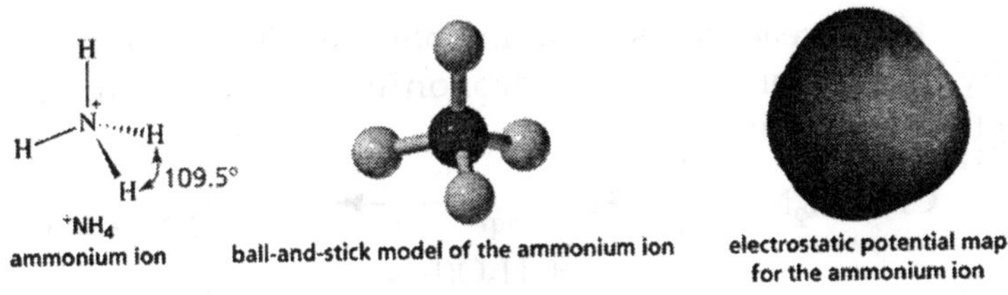

Fig. Ammonium ion

If the reaction is in solution in water (using a dilute acid), the ammonia takes a hydrogen ion (a proton) from a hydroxonium ion. (Remember that hydrogen ions present in solutions of acids in water are carried on water molecules as hydroxonium ions, H_3O^+.)

$$NH_{3(aq)} + H_3O^+_{(aq)} \longrightarrow NH_4^+{}_{(aq)} + H_2O_{(l)}$$

If the acid was hydrochloric acid, for example, you would end up with a solution containing ammonium chloride - the chloride ions, of course, coming from the hydrochloric acid.

You could also write this last equation as:

$$NH_{3(aq)} + H^+_{(aq)} \longrightarrow NH_4^+{}_{(aq)}$$

but if you do it this way, you must include the state symbols. If you write H^+ on its own, it implies an unattached hydrogen ion - a proton. Such things don't exist on their own in solution in water.

If the reaction is happening in the gas state, the ammonia accepts a proton directly from the hydrogen chloride:

$$NH_{3(g)} + HCl_{(g)} \longrightarrow NH_4^+{}_{(3)} + Cl^-{}_{(s)}$$

This time you produce clouds of white solid ammonium chloride.

The Corresponding Reactions with Amines

The nitrogen lone pair behaves exactly the same. The fact that one (or more) of the hydrogens in the ammonia has been replaced by a hydrocarbon group makes no difference.

For example, with ethylamine:

If the reaction is done in solution, the amine takes a hydrogen ion from a hydroxonium ion and forms an ethylammonium ion.

$$CH_3CH_2NH_{2(aq)} + H_3O^+_{(aq)} \longrightarrow CH_3CH_2NH_3^+{}_{(aq)} + H_2O(l)$$

Or:

$$CH_3CH_2NH_{2(aq)} + H^+_{(aq)} \longrightarrow CH_3CH_2NH_3^+{}_{(aq)}$$

The solution would contain ethylammonium chloride or sulphate or whatever.

Alternatively, the amine will react with hydrogen chloride in the gas state to produce the same sort of white smoke as ammonia did - but this time of ethylammonium chloride.

$$CH_3CH_2NH_{2(g)} + HCl_{(g)} \longrightarrow CH_3CH_2NH_3{}^+{}_{(s)} + Cl^-{}_{(s)}$$

These examples have involved a primary amine. It would make no real difference if you used a secondary or tertiary one. The equations would just look more complicated.

The product ions from diethylamine and triethylamine would be diethylammonium ions and triethylammonium ions respectively.

$(CH_3CH_2)_2NH_2{}^+$	$(CH_3CH_2)_3NH^+$
diethylammonium ions	triethylammonium ions

REACTIONS OF AMINES WITH WATER

Again, it is easiest to use the Bronsted-Lowry theory and, again, it is useful to do a straight comparison with ammonia.

A reminder about the Ammonia Reaction with Water

Ammonia is a weak base and takes a hydrogen ion from a water molecule to produce ammonium ions and hydroxide ions.

However, the ammonia is only a weak base, and doesn't hang on to the hydrogen ion very successfully. The reaction is reversible, with the great majority of the ammonia at any one time present as free ammonia rather than ammonium ions.

$$NH_{3(aq)} + H_2O_{(l)} \rightleftharpoons NH_4{}^+{}_{(aq)} + OH^-{}_{(aq)}$$

The presence of the hydroxide ions from this reaction makes the solution alkaline.

The Corresponding Reaction with Amines

The amine still contains the nitrogen lone pair, and does exactly the same thing.

For example, with ethylamine, you get ethylammonium ions and hydroxide ions produced.

$$CH_3CH_2NH_{2(aq)} + H_2O_{(l)} \rightleftharpoons CH_3CH_2NH_3^+{}_{(aq)} + OH^-{}_{(aq)}$$

There is, however, a difference in the position of equilibrium. Amines are usually stronger bases than ammonia. (There are exceptions to this, though - particularly if the amine group is attached directly to a benzene ring.)

REACTIONS OF AMINES WITH COPPER(II) IONS

Just like ammonia, amines react with copper(II) ions in two separate stages. In the first step, we can go on using the Bronsted-Lowry theory (that a base is a hydrogen ion acceptor). The second stage of the reaction can only be explained in terms of the Lewis theory (that a base is an electron pair donor).

The Reaction between Ammonia and Copper(II) ions

Copper(II) sulphate solution, for example, contains the blue hexaaquacopper(II) ion - $[Cu(H_2O)_6]^{2+}$.

In the first stage of the reaction, the ammonia acts as a Bronsted-Lowry base. With a small amount of ammonia solution, hydrogen ions are pulled off two water molecules in the hexaaqua ion.

This produces a neutral complex - one carrying no charge. If you remove two positively charged hydrogen ions from a 2+ ion, then obviously there isn't going to be any charge left on the ion.

Because of the lack of charge, the neutral complex isn't soluble in water, and so you get a pale blue precipitate.

$$[Cu(H_2O)_6]^{2+} + 2NH_3 \rightleftharpoons [Cu(H_2O)_4(OH)_2] + 2NH_4^+$$

This precipitate is often written as $Cu(OH)_2$ and called copper(II) hydroxide. The reaction is reversible because ammonia is only a weak base.

That precipitate dissolves if you add an excess of ammonia solution, giving a deep blue solution.

The ammonia replaces four of the water molecules around the copper to give tetraamminediaquacopper(II) ions. The ammonia uses its lone pair to form a co-ordinate covalent bond (dative covalent bond) with the copper. It is acting as an electron pair donor - a Lewis base.

$$[Cu(H_2O)_6]^{2+} + 4NH_3 \rightleftharpoons [Cu(NH_3)_4(H_2O)_2]_2^{+} + 4H_2O$$

The colour changes are:

The Corresponding Reaction with Amines

The small primary amines behave in exactly the same way as ammonia. There will, however, be slight differences in the shades of blue that you get during the reactions.

Taking methylamine as an example:

With a small amount of methylamine solution you will get a pale blue precipitate of the same neutral complex as with ammonia. All that is happening is that the methylamine is pulling hydrogen ions off the attached water molecules.

$$[Cu(H_2O)_6]^{2+} + 2CH_3NH_2 \rightleftharpoons [Cu(H_2O)_4(OH)_2] + 2CH_3NH_3^{+}$$

With more methylamine solution the precipitate redissolves to give a deep blue solution - just as in the ammonia case. The amine replaces four of the water molecules around the copper.

$$[Cu(H_2O)_6]^{2+} + 4CH_3NH_2 \rightleftharpoons [Cu(CH_3NH_2)_4(H_2O)_2]^{2+} + 4H_2O$$

As the amines get bigger and more bulky, the formula of the final product may change - simply because it is impossible to fit four large amine molecules and two water molecules around the copper atom.

AMINES AS NUCLEOPHILES

Why do amines act as Nucleophiles?

All amines contain an active lone pair of electrons on the very electronegative nitrogen atom. It is these electrons which are attracted to positive parts of other molecules or ions.

REACTIONS OF AMINES

Reactions of Primary Amines with Halogenoalkanes

You get a complicated series of reactions on heating to give a mixture of products - probably one of the most confusing sets of reactions you will meet at this level. The products of the reactions include secondary and tertiary amines and their salts, and quaternary ammonium salts.

Making Secondary Amines and their Salts

In the first stage of the reaction, you get the salt of a secondary amine formed. For example if you started with ethylamine and bromoethane, you would get diethylammonium bromide

$$CH_3CH_2Br + CH_3CH_2NH_2 \longrightarrow (CH_3CH_2)_2NH_2^+ Br^-$$

In the presence of excess ethylamine in the mixture, there is the possibility of a reversible reaction. The ethylamine removes a hydrogen from the diethylammonium ion to give free diethylamine - a secondary amine.

$$(CH_3CH_2)_2NH_2^+ Br^- + CH_3CH_2NH_2 \rightleftharpoons (CH_3CH_2)_2NH + CH_3CH_2NH_3^+ Br^-$$

Making Tertiary Amines and their Salts

But it doesn't stop here! The diethylamine also reacts with bromoethane - in the same two stages as before. This is where the reaction would start if you reacted a secondary amine with a halogenoalkane.

In the first stage, you get triethylammonium bromide.

$$CH_3CH_2Br + (CH_3CH_2)_2NH \longrightarrow (CH_3CH_2)_3NH^+ Br^-$$

There is again the possibility of a reversible reaction between this salt and excess ethylamine in the mixture.

$$(CH_3CH_2)_2\overset{+}{N}H(CH_2CH_3)\,Br^- + CH_3CH_2NH_2 \rightleftharpoons CH_3CH_2-N(CH_2CH_3)_2 + CH_3CH_2NH_3^+\,Br^-$$

The ethylamine removes a hydrogen ion from the triethylammonium ion to leave a tertiary amine - triethylamine.

Making a Quaternary Ammonium Salt

The final stage! The triethylamine reacts with bromoethane to give tetraethylammonium bromide - a quaternary ammonium salt (one in which all four hydrogens have been replaced by alkyl groups).

$$CH_3CH_2Br + CH_3CH_2-N(CH_2CH_3)_2 \longrightarrow CH_3CH_2-\overset{+}{N}(CH_2CH_3)_2-CH_2CH_3\ Br^-$$

This time there isn't any hydrogen left on the nitrogen to be removed. The reaction stops here.

REACTIONS WITH ACYL CHLORIDES (ACID CHLORIDES)

We'll take the reaction between methylamine and ethanoyl chloride as typical.

If you add concentrated methylamine solution to ethanoyl chloride, there is a violent reaction in the cold. N-methylethanamide and methylammonium chloride are formed - partly as a white solid mixture, and partly in solution.

The overall equation is:

$$CH_3COCl + 2CH_3NH_2 \longrightarrow CH_3CONHCH_3 + CH_3NH_3Cl$$

REACTIONS WITH ACID ANHYDRIDES

These reactions are chemically similar to those between amines and acyl chlorides, but they are much slower, needing heat.

Taking the reaction between methylamine and ethanoic anhydride as typical:

The product is N-methylethanamide (as with ethanoyl chloride), but this time the other product is methylammonium ethanoate rather than methylammonium chloride.

$$(CH_3CO)_2O + 2CH_3NH_2 \longrightarrow CH_3CONHCH_3 + CH_3COO^- + NH_3CH_3$$

REACTION WITH NITROUS ACID

Testing for the Various types of Amines

The reaction between amines and nitrous acid was used in the past as a very neat way of distinguishing between primary, secondary and tertiary amines. However, the product with a secondary amine is a powerful carcinogen, and so this reaction is no longer carried out at this level.

Nitrous acid, HNO_2, (sometimes written as HONO to show its structure) is unstable and is always prepared *in situ.*

It is usually made by reacting a solution containing sodium or potassium nitrite (sodium or potassium nitrate(III)) with hydrochloric acid.

Nitrous acid is a weak acid and so you get the reaction:

$$H^+_{(aq)} + NO_2^-{}_{(aq)} \rightleftharpoons HNO_{2(aq)}$$

Because nitrous acid is a weak acid, the position of equilibrium lies well the right.

In each of the following reactions, the amine would be acidified with hydrochloric acid and a solution of sodium or potassium nitrite added. The acid and the nitrite form nitrous acid which then reacts with the amine.

Primary Amines and Nitrous Acid

The main observation is a burst of colourless, odourless gas. Nitrogen is given off.

Unfortunately, there is no single clear-cut equation that you can quote for this. You get lots of different organic products. For example, amongst the products you get an alcohol where the -NH_2 group has been replaced by OH. If you want a single equation, you could quote (taking 1-aminopropane as an example):

$$CH_3CH_2CH_2NH_2 + HNO_2 \longrightarrow CH_3CH_2CH_2OH + H_2O + N_2$$

but the propan-1-ol will be only one product among many - including propan-2-ol, propene, 1-chloropropane, 2-chloropropane and others.

The nitrogen, however, is given off in quantities exactly as suggested by the equation. By measuring the amount of nitrogen produced, you could use this reaction to work out the amount of amine present in the solution.

Secondary Amines and Nitrous Acid

This time there isn't any gas produced. Instead, you get a yellow oil called a nitrosamine. These compounds are powerful carcinogens.

For example:

$$(CH_3)_2NH + HNO_2 \longrightarrow (CH_3)_2N\text{-}N{=}O + H_2O$$

Tertiary Amines and Nitrous Acid

Again, a quite different result. This time, nothing visually interesting happens - you are left with a colourless solution.

All that has happened is that the amine has formed an ion by reacting with the acid present. With trimethylamine, for example, you would get a trimethylammonium ion, $(CH_3)_3NH^+$.

Chapter 15

Nitriles

INTRODUCTION

Nitriles contain the -CN group, and used to be known as cyanides.

Some Simple Nitriles

The smallest organic nitrile is ethanenitrile, CH_3CN, (old name: methyl cyanide or acetonitrile - and sometimes now called ethanonitrile). Hydrogen cyanide, HCN, doesn't usually count as organic, even though it contains a carbon atom.

$$CH_3-C\equiv N$$

ethanenitrile

Notice the triple bond between the carbon and nitrogen in the -CN group.

The three simplest nitriles are:

CH_3CN	Ethanenitrile
CH_3CH_2CN	Propanenitrile
$CH_3CH_2CH_2CN$	Butanenitrile

When you are counting the length of the carbon chain, don't forget the carbon in the -CN group. If the chain is branched, this carbon usually counts as the number 1 carbon.

PREPARATION

From Halogenoalkanes

The halogenoalkane is heated under reflux with a solution of sodium or potassium cyanide in ethanol. The halogen is replaced by a -CN group and a nitrile is produced. Heating under reflux means heating with a condenser placed vertically in the flask to prevent loss of volatile substances from the mixture.

The solvent is important. If water is present you tend to get substitution by -OH instead of -CN.

For example, using 1-bromopropane as a typical halogenoalkane:

$$CH_3CH_2CH_2Br + CN^- \longrightarrow CH_3CH_2CH_2CN + Br^-$$

You could write the full equation rather than the ionic one, but it slightly obscures what's going on:

$$CH_3CH_2CH_2Br + KCN \longrightarrow CH_3CH_2CH_2CN + KBr$$

The bromine (or other halogen) in the halogenoalkane is simply replaced by a -CN group - hence a substitution reaction. In this example, butanenitrile is formed.

Making a nitrile by this method is a useful way of increasing the length of a carbon chain. Having made the nitrile, the -CN group can easily be modified to make other things.

From Amides

Nitriles can be made by dehydrating amides.

Amides are dehydrated by heating a solid mixture of the amide and phosphorus(V) oxide, P_4O_{10}.

Water is removed from the amide group to leave a nitrile group, -CN. The liquid nitrile is collected by simple distillation.

For example, you will get ethanenitrile by dehydrating ethanamide.

$$CH_3CONH_2 \xrightarrow[-H_2O]{P_2O_{10}} CH_3CN$$

From Aldehydes and Ketones

Aldehydes and ketones undergo an addition reaction with hydrogen cyanide. The hydrogen cyanide adds across the carbon-oxygen double bond in the aldehyde or ketone to produce a hydroxynitrile. Hydroxynitriles used to be known as cyanohydrins.

For example, with ethanal (an aldehyde) you get 2-hydroxypropanenitrile:

$$CH_3-C(=O)H + HCN \longrightarrow CH_3-C(OH)(H)-CN$$

With propanone (a ketone) you get 2-hydroxy-2-methylpropanenitrile:

$$CH_3-C(=O)H + HCN \longrightarrow CH_3-C(OH)(H)-CN$$

In every example of this kind, the -OH group will be on the number 2 carbon atom - the one next to the -CN group.

The reaction isn't normally done using hydrogen cyanide itself, because this is an extremely poisonous gas. Instead, the aldehyde or ketone is mixed with a solution of sodium or potassium cyanide in water to which a little sulphuric acid has been added. The pH of the solution is adjusted to about 4-5, because this gives the fastest reaction. The reaction happens at room temperature.

The solution will contain hydrogen cyanide (from the reaction between the sodium or potassium cyanide and the sulphuric acid), but still contains some free cyanide ions. This is important for the mechanism.

These are useful reactions because they not only increase the number of carbon atoms in a chain, but also introduce another reactive group as well as the -CN group. The -OH group behaves just like the -OH group in any alcohol with a similar structure.

For example, starting from a hydroxynitrile made from an aldehyde, you can quite easily produce relatively complicated molecules like 2-amino acids - the amino acids which are used to construct proteins.

$$R\text{-}\underset{H}{\overset{OH}{C}}\text{-}CN \longrightarrow R\text{-}\underset{H}{\overset{Cl}{C}}\text{-}CN \longrightarrow R\text{-}\underset{H}{\overset{NH_2}{C}}\text{-}CN \longrightarrow R\text{-}\underset{H}{\overset{NH_2}{C}}\text{-}COOH$$

PHYSICAL PROPERTIES

Boiling Points

The small nitriles are liquids at room temperature.

Nitrile	Boiling Point (°C)
CH_3CN	82
CH_3CH_2CN	97
$CH_3CH_2CH_2CN$	116-118

These boiling points are very high for the size of the molecules - similar to what you would expect if they were capable of forming hydrogen bonds.

However, they *don't* form hydrogen bonds - they don't have a hydrogen atom directly attached to an electronegative element.

They are just very polar molecules. The nitrogen is very electronegative and the electrons in the triple bond are very easily pulled towards the nitrogen end of the bond.

Nitriles therefore have strong permanent dipole-dipole attractions as well as Van der Waals dispersion forces between their molecules.

Solubility in Water

Ethanenitrile is completely soluble in water, and the solubility then falls as chain length increases.

Nitrile	Solubility at 20°C
CH_3CN	Miscible
CH_3CH_2CN	10 g per 100 cm3 of water
$CH_3CH_2CH_2CN$	3 g per 100 cm3 of water

The reason for the solubility is that although nitriles can't hydrogen bond with themselves, they can hydrogen bond with water molecules.

One of the slightly positive hydrogen atoms in a water molecule is attracted to the lone pair on the nitrogen atom in a nitrile and a hydrogen bond is formed.

$$CH_3-C\equiv N^{\delta-}\!:\cdots H^{\delta+}-O-H$$

hydrogen bond

There will also, of course, be dispersion forces and dipole-dipole attractions between the nitrile and water molecules.

Forming these attractions releases energy. This helps to supply the energy needed to separate water molecule from water molecule and nitrile molecule from nitrile molecule before they can mix together.

As chain lengths increase, the hydrocarbon parts of the nitrile molecules start to get in the way.

By forcing themselves between water molecules, they break the relatively strong hydrogen bonds between water molecules without replacing them by anything as good. This makes the process energetically less profitable, and so solubility decreases.

HYDROLYSIS

When nitriles are hydrolyzed you can think of them reacting with water in two stages - first to produce an amide, and then the ammonium salt of a carboxylic acid.

For example, ethanenitrile would end up as ammonium ethanoate going via ethanamide.

$$CH_3CN \xrightarrow{H_2O} CH_3CONH_2 \xrightarrow{H_2O} CH_3COONH_4$$

In practice, the reaction between nitriles and water would be so slow as to be completely negligible. The nitrile is instead heated with either a dilute acid such as dilute hydrochloric acid, or with an alkali such as sodium hydroxide solution.

The end result is similar in all the cases, but the exact nature of the final product varies depending on the conditions you use for the reaction.

Acidic Hydrolysis of Nitriles

The nitrile is heated under reflux with dilute hydrochloric acid. Instead of getting an ammonium salt as you would do if the reaction only involved water, you produce the free carboxylic acid.

For example, with ethanenitrile and hydrochloric acid you would get ethanoic acid and ammonium chloride.

$$CH_3CN + 2H_2O + HCl \longrightarrow CH_3COOH + NH_4Cl$$

Why is the free acid formed rather than the ammonium salt? The ethanoate ions in the ammonium ethanoate react with hydrogen ions from the hydrochloric acid to produce ethanoic acid. Ethanoic acid is only a weak acid and so once it has got the hydrogen ion, it tends to hang on to it.

Alkaline Hydrolysis of Nitriles

The nitrile is heated under reflux with sodium hydroxide solution. This time, instead of getting an ammonium salt as you would do if the reaction only involved water, you get the sodium salt. Ammonia gas is given off as well.

For example, with ethanenitrile and sodium hydroxide solution you would get sodium ethanoate and ammonia.

$$CH_3CN + H_2O + NaOH \longrightarrow CH_3COONa + NH_3$$

The ammonia is formed from reaction between ammonium ions and hydroxide ions.

If you wanted the free carboxylic acid in this case, you would have to acidify the final solution with a strong acid such as dilute hydrochloric acid or dilute sulphuric acid. The ethanoate ion in the sodium ethanoate will react with hydrogen ions as mentioned above.

REDUCTION OF NITRILES TO PRIMARY AMINES

Reduction of Nitriles using $LiAlH_4$

The structure of $LiAlH_4$ is:

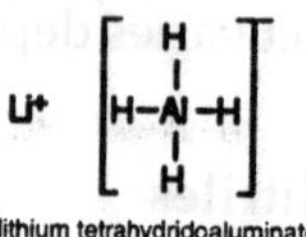

lithium tetrahydridoaluminate

In the negative ion, one of the bonds is a co-ordinate covalent (dative covalent) bond using the lone pair on a hydride ion (H^-) to form a bond with an empty orbital on the aluminium.

Overall Reaction

The nitrile reacts with the lithium tetrahydridoaluminate in solution in ethoxyethane (diethyl ether, or just "ether") followed by treatment of the product of that reaction with a dilute acid.

Overall, the carbon-nitrogen triple bond is reduced to give a primary amine. Primary amines contain the $-NH_2$ group.

For example, with ethanenitrile you get ethylamine:

$$CH_3CN + 2[H] \longrightarrow CH_3CH_2NH_2$$

Notice that this is a simplified equation. [H] means "hydrogen from a reducing agent".

REDUCTION OF NITRILES USING HYDROGEN AND A METAL CATALYST

The carbon-nitrogen triple bond in a nitrile can also be reduced by reaction with hydrogen gas in the presence of a variety of metal catalysts.

Commonly quoted catalysts are palladium, platinum or nickel.

The reaction will take place at a raised temperature and pressure. It is impossible to give exact details because it will vary from catalyst to catalyst.

For example, ethanenitrile can be reduced to ethylamine by reaction with hydrogen in the presence of a palladium catalyst.

$$CH_3CN + 2H_2 \xrightarrow{Pd} CH_3CH_2NH_2$$

Chapter 16

Arenes or Aromatic Hydrocarbons

INTRODUCTION

Arenes are aromatic hydrocarbons. The term "aromatic" originally referred to their pleasant smells, but now implies a particular sort of delocalized bonding.

The arenes are based on benzene rings. The simplest of them is benzene itself, C_6H_6. The next simplest is methylbenzene (old name: toluene) which has one of the hydrogen atoms attached to the ring replaced by a methyl group - $C_6H_5CH_3$.

AROMATICITY

Aromaticity is a chemical property in which a conjugated ring of unsaturated bonds, lone pairs, or empty orbitals exhibit a stabilization stronger than would be expected by the stabilization of conjugation alone. It can also be considered a manifestation of cyclic delocalization and of resonance. This is usually considered to be because electrons are free to cycle around circular arrangements of atoms, which are alternately single- and double-bonded to one another. These bonds may be seen as a hybrid of a single bond and a double bond, each bond in the ring identical to every other. This commonly-seen model of aromatic rings was developed by Kekule. The model for benzene consists of two resonance forms, which corresponds to the double and single bonds' switching positions. Benzene is a more stable molecule than would be expected without accounting for charge delocalization.

THEORY

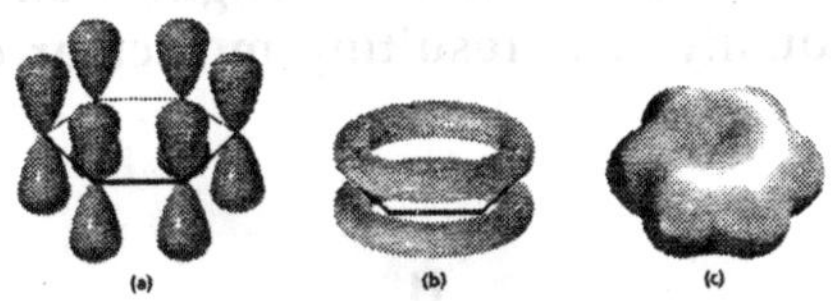

Fig. Benzene molecule

By convention, the double-headed arrow indicates that the two structures are simply hypothetical, since neither is an accurate representation of the actual compound. The actual molecule is best represented by a hybrid (average) of these structures, which can be seen above. A C–C bond is shorter than a C=C bond, but benzene is perfectly hexagonal—all six carbon-carbon bonds have the same length, intermediate between that of a single and that of a double bond.

A better representation is that of the circular Π bond (Armstrong's *inner cycle*), in which the electron density is evenly distributed through a Π bond above and below the ring. This model more correctly represents the location of electron density within the aromatic ring.

The single bonds are formed with electrons in line between the carbon nuclei—these are called sigma bonds. Double bonds consist of a sigma bond and another bond—a Π bond. The Π-bonds are formed from overlap of atomic p-orbitals above and below the plane of the ring. The following diagram shows the positions of these p-orbitals:

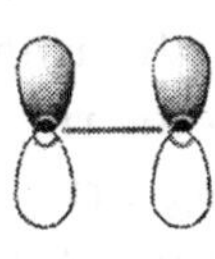

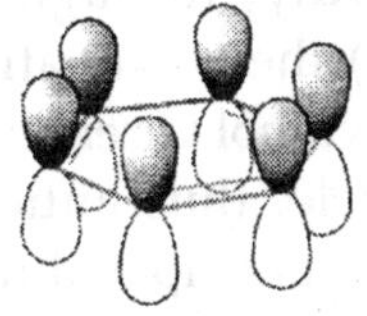

Side View Projection

Since they are out of the plane of the atoms, these orbitals can interact with each other freely, and become delocalized.

This means that instead of being tied to one atom of carbon, each electron is shared by all six in the ring. Thus, there are not enough electrons to form double bonds on all the carbon atoms, but the "extra" electrons strengthen all of the bonds on the ring equally. The resulting molecular orbital has P symmetry.

HISTORY

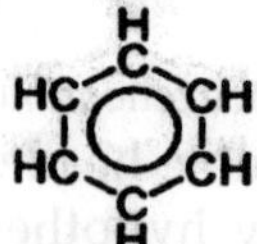

The attribution of this exceptional stability is conventionally to Sir Robert Robinson, who was apparently the first (in 1925) to coin the term aromatic sextet as a group of six electrons that resists disruption.

In fact, this concept can be traced further back, via Ernest Crocker in 1922, to Henry Edward Armstrong, who in 1890, in an article entitled The structure of cycloid hydrocarbons, wrote the (six) centric affinities act within a cycle...benzene may be represented by a double ring (sic) ... and when an additive compound is formed, the inner cycle of affinity suffers disruption, the contiguous carbon-atoms to which nothing has been attached of necessity acquire the ethylenic condition.

Here he is describing at least four modern concepts. Firstly his affinity is better known nowadays as the electron, which was only to be discovered seven years later by J. J. Thomson. Secondly, he is describing electrophilic aromatic substitution proceeding (thirdly) through a Wheland intermediate, in which (fourthly) the conjugation of the ring is broken. He introduced the symbol C centered on the ring as a shorthand for the inner cycle, thus anticipating Eric Clar's notation. Arguably, he also anticipated the nature of wave mechanics, since he recognized that his affinities had direction, not merely being point particles, and collectively having a distribution that could be altered by introducing substituents onto the benzene ring (much as the distribution of the electric charge in a body is altered by bringing it near to another body). The quantum mechanical origins of this stability, or aromaticity, were first modelled by Hückel in 1931.

CHARACTERISTICS OF AROMATIC COMPOUNDS

An aromatic compound contains a set of covalently-bound atoms with specific characteristics:

1. A delocalized conjugated Π system, most commonly an arrangement of alternating single and double bonds
2. Coplanar structure, with all the contributing atoms in the same plane
3. Contributing atoms arranged in one or more rings
4. A number of Π delocalized electrons that is even, but not a multiple of 4. This is known as Hückel's rule. Permissible numbers of Π electrons include 2, 6, 10, 14, and so on
5. Special reactivity in organic reactions such as electrophilic aromatic substitution and nucleophilic aromatic substitution

Whereas benzene is aromatic (6 electrons, from 3 double bonds), cyclobutadiene is not, since the number of P delocalized electron is 4, which of course is a multiple of 4. The cyclobutadienide (2Π) ion, however, is aromatic (6 electrons). An atom in an aromatic system can have other electrons that are not part of the system, and are therefore ignored for the 4n + 2 rule. In furan, the oxygen atom is sp^2 hybridized. One lone pair is in the Π system and the other in the plane of the ring (analogous to C-H bond on the other positions). There are 6 Π electrons, so furan is aromatic.

Aromatic molecules typically display enhanced chemical stability, compared to similar non-aromatic molecules. The circulating Π electrons in an aromatic molecule generate significant local magnetic fields that can be detected by NMR techniques. NMR experiments show that protons on the aromatic ring are shifted substantially further down-field than those on aliphatic carbons. Planar monocyclic molecules containing 4n Π electrons are called antiaromatic and are, in general, destabilized. Molecules that could be antiaromatic will tend to alter their electronic or conformational structure to avoid this situation, thereby becoming non-aromatic. For example, cyclooctatetraene (COT) distorts itself out of planarity, breaking Πoverlap between adjacent double bonds. Möbius aromaticity describes a special case of aromaticity.

Aromatic molecules are able to interact with each other in so-called Π-Π stacking: the Π systems form two parallel rings overlap in a "face-to-face" orientation. Aromatic molecules are also able to interact with each other in an "edge-to-face" orientation: the slight positive charge of the substituents on the ring atoms of one molecule are attracted to the slight negative charge of the aromatic system on another molecule.

Many of the earliest-known examples of aromatic compounds, such as benzene and toluene, have distinctive pleasant smells. This property led to the term "aromatic" for this class of compounds, and hence to "aromaticity" being the eventually-discovered electronic property of them.

Structure of Benzene

What you need to understand about benzene is:

- Benzene, C_6H_6, is a planar molecule containing a ring of six carbon atoms each with a hydrogen atom attached.

  The six carbon atoms form a perfectly regular hexagon. All the carbon-carbon bonds have exactly the same lengths - somewhere between single and double bonds.

- There are delocalized electrons above and below the plane of the ring.

The Benzene Ring

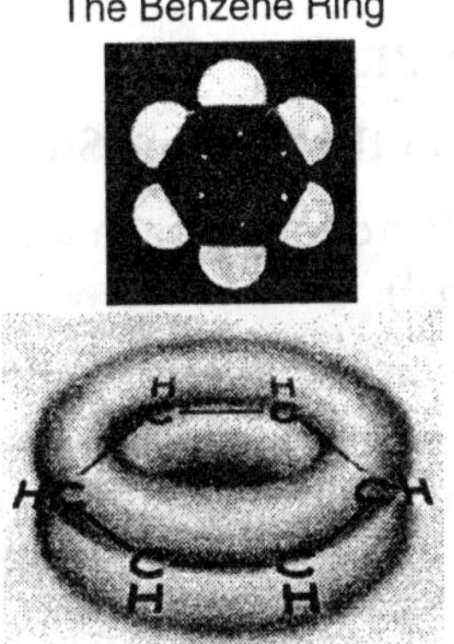

Fig. Structure of Benzene

This diagram shows one of the molecular orbitals containing two of the delocalized electrons. The other molecular orbitals are almost never drawn.

- The presence of the delocalized electrons makes benzene particularly stable.
- Benzene resists addition reactions because that would involve breaking the delocalization and losing that stability.

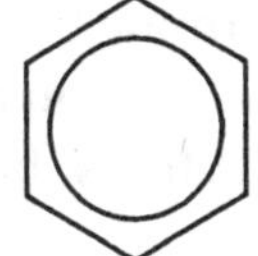

Benzene is represented by this symbol, where the circle represents the delocalized electrons, and each corner of the hexagon has a carbon atom with a hydrogen attached.

The Structure of Methylbenzene (Toluene)

Methylbenzene just has a methyl group attached to the benzene ring - replacing one of the hydrogen atoms.

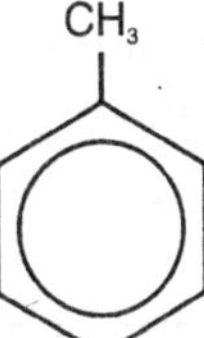

Attached groups are often drawn at the top of the ring, but you may occasionally find them drawn in other places with the ring rotated.

BONDING IN BENZENE

An orbital Model for the Benzene Structure

Benzene is built from hydrogen atoms ($1s^1$) and carbon atoms ($1s^2 2s^2 2p_x{}^1 2p_y{}^1$).

Each carbon atom has to join to three other atoms (one hydrogen and two carbons) and doesn't have enough unpaired electrons to form the required number of bonds, so it needs to promote one of the $2s^2$ pair into the empty $2p_z$ orbital. This is exactly the same as happens whenever carbon forms bonds - whatever else it ends up joined to.

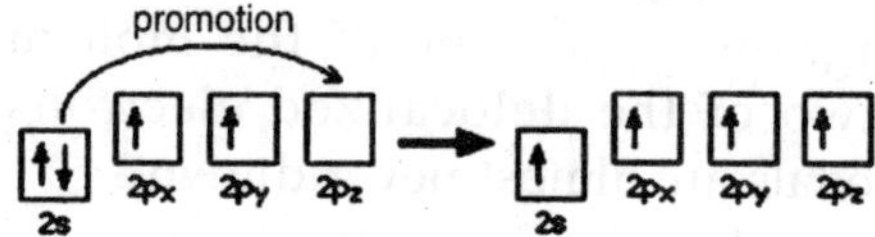

Because each carbon is only joining to three other atoms, when the carbon atoms hybridise their outer orbitals before forming bonds, they only need to hybridise three of the orbitals rather than all four. They use the 2s electron and two of the 2p electrons, but leave the other 2p electron unchanged.

The new orbitals formed are called sp^2 hybrids, because they are made by an s orbital and two p orbitals reorganising themselves. The three sp^2 hybrid orbitals arrange themselves as far apart as possible - which is at 120° to each other in a plane. The remaining p orbital is at right angles to them.

The p electron on each carbon atom overlaps with those on both sides of it. This extensive sideways overlap produces a system of pi bonds which are spread out over the whole carbon ring. Because the electrons are no longer held between just two carbon atoms, but are spread over the whole ring, the electrons are said to be delocalized. The six delocalized electrons go into three molecular orbitals - two in each.

In common with the great majority of descriptions of the bonding in benzene, we are only going to show one of these delocalized molecular orbitals for simplicity.

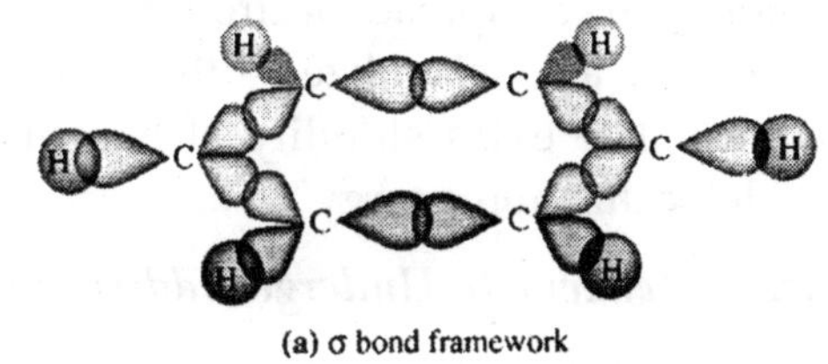

(a) σ bond framework

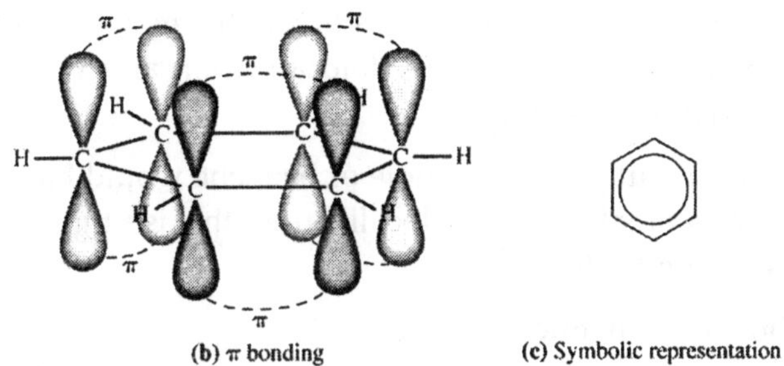

(b) π bonding

(c) Symbolic representation

Fig. Bonding in Benzene

In the diagram, the sigma bonds have been shown as simple lines to make the diagram less confusing. The two rings above and below the plane of the molecule represent *one* molecular orbital. The two delocalized electrons can be found anywhere within those rings. The other four delocalized electrons live in two similar (but not identical) molecular orbitals.

Relating the Orbital Model to the Properties of Benzene

The Shape of Benzene

This is easily explained. Benzene is a regular hexagon because all the bonds are identical. The delocalization of the electrons means that there aren't alternating double and single bonds.

The Energetic Stability of Benzene

This is accounted for by the delocalization. As a general principle, the more you can spread electrons around - in other words, the more they are delocalized - the more stable the molecule becomes. The extra stability of benzene is often referred to as "delocalization energy".

The Reluctance of Benzene to Undergo Addition Reactions

With the delocalized electrons in place, benzene is about 150 kJ mol^{-1} more stable than it would otherwise be. If you added other atoms to a benzene ring you would have to use some of the delocalized electrons to join the new atoms to the ring. That would disrupt the delocalization and the system would become less stable.

Since about 150 kJ per mole of benzene would have to be supplied to break up the delocalization, this isn't going to be an easy thing to do.

Symbol for Benzene

Although you will still come across the Kekulé structure for benzene, for most purposes we use the structure below.

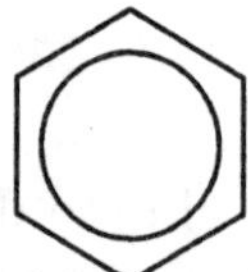

The hexagon shows the ring of six carbon atoms, each of which has one hydrogen attached. The circle represents the delocalized electrons. It is essential that you include the circle. If you miss it out, you are drawing cyclohexane and not benzene.

BONDING IN BENZENE

The Kekule Structure for Benzene, C_6H_6

Kekulé was the first to suggest a sensible structure for benzene. The carbons are arranged in a hexagon, and he suggested alternating double and single bonds between them. Each carbon atom has a hydrogen attached to it.

This diagram is often simplified by leaving out all the carbon and hydrogen atoms!

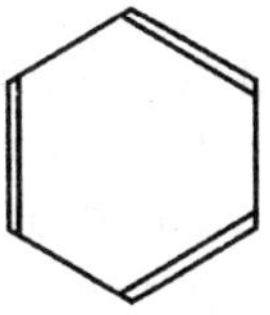

In diagrams of this sort, there is a carbon atom at each corner. You have to count the bonds leaving each carbon to work out how many hydrogens there are attached to it.

In this case, each carbon has three bonds leaving it. Because carbon atoms form four bonds, that means you are a bond missing - and that must be attached to a hydrogen atom.

Problems with the Kekule Structure

Although the Kekulé structure was a good attempt in its time, there are serious problems with it.

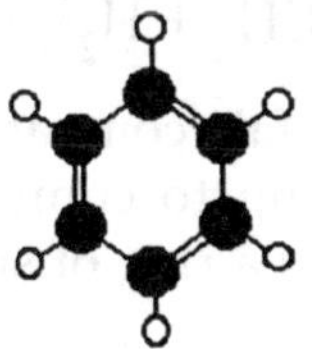

Problems with the Chemistry

Because of the three double bonds, you might expect benzene to have reactions like ethane. Ethene undergoes addition reactions in which one of the two bonds joining the carbon atoms breaks, and the electrons are used to bond with additional atoms.

Benzene rarely does this. Instead, it usually undergoes substitution reactions in which one of the hydrogen atoms is replaced by something new.

Problems with the Shape

Benzene is a planar molecule (all the atoms lie in one plane), and that would also be true of the Kekulé structure. The problem is that C-C single and double bonds are different lengths.

C-C	0.154 nm
C=C	0.134 nm

That would mean that the hexagon would be irregular if it had the Kekulé structure, with alternating shorter and longer sides. In real benzene all the bonds are exactly the same - intermediate in length between C-C and C=C at 0.139 nm. Real benzene is a perfectly regular hexagon.

Problems with the Stability of Benzene

Real benzene is a lot more stable than the Kekulé structure would give it credit for. Every time you do a thermochemistry calculation based on the Kekulé structure, you get an answer which is wrong by about 150 kJ mol^{-1}. This is most easily shown using enthalpy changes of hydrogenation.

Hydrogenation is the addition of hydrogen to something. If, for example, you hydrogenate ethene you get ethane:

$$CH_2{=}CH_2 + H_2 \longrightarrow CH_3CH_3$$

In order to do a fair comparison with benzene (a ring structure) we're going to compare it with cyclohexene. Cyclohexene, C_6H_{10}, is a ring of six carbon atoms containing just one C=C.

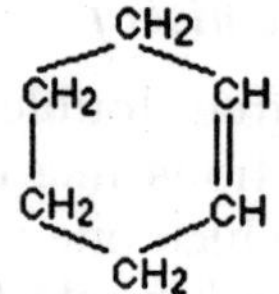

When hydrogen is added to this, cyclohexane, C_6H_{12}, is formed. The "CH" groups become CH_2 and the double bond is replaced by a single one.

The structures of cyclohexene and cyclohexane are usually simplified in the same way that the Kekulé structure for benzene is simplified - by leaving out all the carbons and hydrogens.

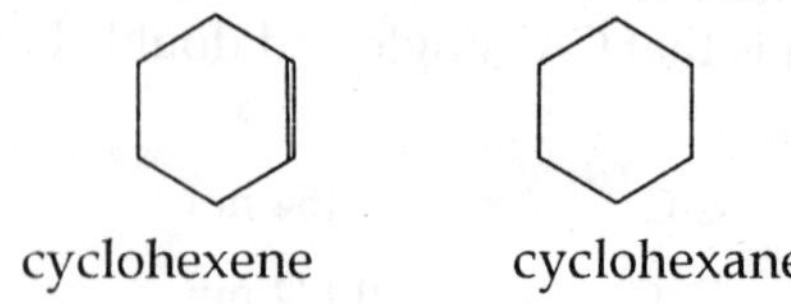

cyclohexene cyclohexane

In the cyclohexane case, for example, there is a carbon atom at each corner, and enough hydrogens to make the total bonds on each carbon atom up to four. In this case, then, each corner represents CH_2.

The hydrogenation equation could be written:

The enthalpy change during this reaction is -120 kJ mol^{-1}. In other words, when 1 mole of cyclohexene reacts, 120 kJ of heat energy is evolved.

Where does this heat energy come from? When the reaction happens, bonds are broken (C=C and H-H) and this costs energy. Other bonds have to be made, and this releases energy.

Because the bonds made are stronger than those broken, more energy is released than was used to break the original bonds and so there is a net evolution of heat energy.

If the ring had *two* double bonds in it initially (*cyclohexa-1,3-diene*), exactly twice as many bonds would have to be broken and exactly twice as many made. In other words, you would expect the enthalpy change of hydrogenation of cyclohexa-1,3-diene to be exactly twice that of cyclohexene - that is, -240 kJ mol^{-1}.

+ $2H_2$ ⟶

In fact, the enthalpy change is *-232 kJ mol^{-1}* - which isn't far off what we are predicting.

Applying the same argument to the Kekulé structure for benzene (what might be called *cyclohexa-1,3,5-triene*), you would expect an enthalpy change of -360 kJ mol^{-1}, because there are exactly three times as many bonds being broken and made as in the cyclohexene case.

+ $3H_2$ ⟶

In fact what you get is *-208 kJ mol^{-1}* - not even within distance of the predicted value!

This is very much easier to see on an enthalpy diagram. Notice that in each case heat energy is released, and in each case the product is the same (cyclohexane). That means that all the reactions "fall down" to the same end point.

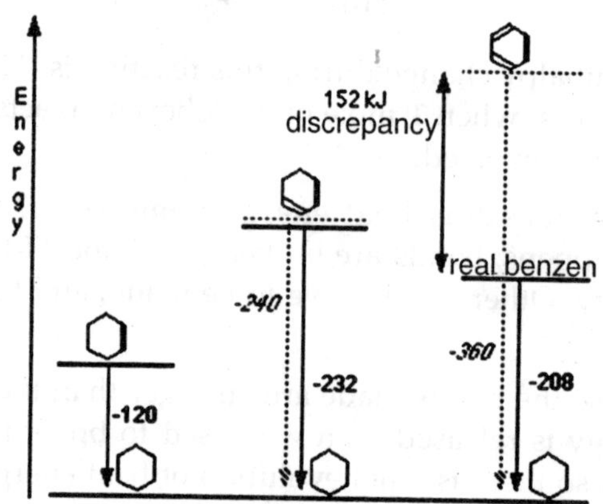

Heavy lines, solid arrows and bold numbers represent real changes. Predicted changes are shown by dotted lines and italics.

The most important point to notice is that real benzene is much lower down the diagram than the Kekulé form predicts. The lower down a substance is, the more energetically stable it is.

This means that real benzene is about 150 kJ mol^{-1} more stable than the Kekulé structure gives it credit for. This increase in stability of benzene is known as the delocalisation energy or resonance energy of benzene. The first term (delocalisation energy) is the more commonly used.

PREPARATION

Here we look at the manufacture of arenes such as benzene and methylbenzene (toluene) by the catalytic reforming of fractions from petroleum (crude oil).

Catalytic Reforming

Reforming takes straight chain hydrocarbons in the C_6 to C_8 range from the gasoline or naphtha fractions and rearranges them into compounds containing benzene rings. Hydrogen is produced as a by-product of the reactions.

Hexane, C_6H_{14}, loses hydrogen and turns into benzene. As long as you draw the hexane bent into a circle, it is easy to see what is happening.

$$CH_3-CH_2-CH_2-CH_2-CH_2-CH_3 \longrightarrow C_6H_6 + 4H_2$$

Similarly, methylbenzene (toluene) is made from heptane:

$$CH_3-CH_2-CH_2-CH_2-CH_2-CH_2-CH_3 \longrightarrow C_6H_5CH_3 + 4H_2$$

PROCESS

From Feedstock

The feedstock is a mixture of the naphtha or gasoline fractions and hydrogen. The hydrogen is there to help prevent the formation of carbon by decomposition of the hydrocarbons at the high temperatures used. The carbon would otherwise contaminate the catalyst.

Catalyst

A typical catalyst is a mixture of platinum and aluminium oxide. With a platinum catalyst, the process is sometimes described as "platforming".

Temperature and Pressure

The temperature is about 500°C, and the pressure varies either side of 20 atmospheres.

Converting some of the Methylbenzene into Benzene

Methylbenzene is much less commercially valuable than benzene. The methyl group can be removed from the ring by a process known as "dealkylation".

The methylbenzene is mixed with hydrogen at a temperature of between 550 and 650°C, and a pressure of between 30 and 50 atmospheres, with a mixture of silicon dioxide and aluminium oxide as catalyst.

$$C_6H_5CH_3 + H_2 \longrightarrow C_6H_6 + CH_4$$

PHYSICAL PROPERTIES

Boiling Points

In benzene, the only attractions between neighbouring molecules are Van der Waals dispersion forces. There is no permanent dipole on the molecule.

Benzene boils at 80°C - rather higher than other hydrocarbons of similar molecular size (pentane and hexane, for example). This is presumably due to the ease with which temporary dipoles can be set up involving the delocalized electrons.

Methylbenzene boils at 111°C. It is a bigger molecule and so the Van der Waals dispersion forces will be bigger.

Methylbenzene also has a small permanent dipole, so there will be dipole-dipole attractions as well as dispersion forces. The dipole is due to the CH_3 group's tendency to "push" electrons away from itself. This also affects the reactivity of methylbenzene.

Melting Points

You might have expected that methylbenzene's melting point would be higher than benzene's as well, but it isn't - it is much lower! Benzene melts at 5.5°C; methylbenzene at -95°C.

Molecules must pack efficiently in the solid if they are to make best use of their intermolecular forces. Benzene is a tidy, symmetrical molecule and packs very efficiently. The methyl group sticking out in methylbenzene tends to disrupt the closeness of the packing. If the molecules aren't as closely packed, the intermolecular forces don't work as well and so the melting point falls.

Solubility in Water

The arenes are insoluble in water.

Benzene is quite large compared with a water molecule. In order for benzene to dissolve it would have to break lots of existing hydrogen bonds between water molecules. You also have to break the quite strong Van der Waals dispersion forces between benzene molecules. Both of these cost energy.

The only new forces between the benzene and the water would be Van der Waals dispersion forces. These aren't as strong as hydrogen bonds (or the original dispersion forces in the benzene), and so you wouldn't get much energy released when they form.

It simply isn't energetically profitable for benzene to dissolve in water. It would, of course, be even worse for larger arene molecules.

REACTIVITY

Benzene

It has already been pointed out above that benzene is resistant to addition reactions. Adding something new to the ring would need you to use some of the delocalized electrons to form bonds with whatever you are adding. That results in a major loss of stability as the delocalization is broken.

Instead, benzene mainly undergoes substitution reactions - replacing one or more of the hydrogen atoms by something new. That leaves the delocalized electrons as they were.

Methylbenzene

You have to consider the reactivity of something like methylbenzene in two distinct bits:

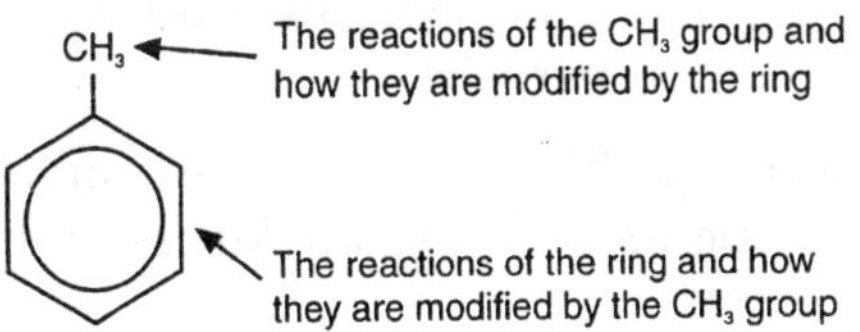

For example, if you explore other pages in this section, you will find that alkyl groups attached to a benzene ring are oxidised by alkaline potassium manganate(VII) solution. This doesn't happen in the absence of the benzene ring.

The tendency of the CH_3 group to "push" electrons away from itself also has an effect on the ring, making methylbenzene react more quickly than benzene itself.

REACTIONS OF BENZENE AND METHYLBENZENE

Here we are going to deal with the combustion, hydrogenation and sulphonation of benzene and methylbenzene (toluene), and with the oxidation of side chains attached to benzene rings.

Combustion

Like any other hydrocarbons, benzene and methylbenzene burn in a plentiful supply of oxygen to give carbon dioxide and water. For example:

For benzene:

$$2C_6H_6 + 15O_2 \rightarrow 12CO_2 + 6H_2O$$

and methylbenzene:

$$C_6H_5CH_3 + 9O_2 \rightarrow 7CO_2 + 4H_2O$$

However, for these hydrocarbons, combustion is hardly ever complete, especially if they are burnt in air. The high proportion of carbon in the molecules means that you need a very high proportion of oxygen to hydrocarbon to get complete combustion. Look at the equations.

As a general rule, the hydrogen in a hydrocarbon tends to get what oxygen is available first, leaving the carbon to form carbon itself, or carbon monoxide, if there isn't enough oxygen to go round.

The arenes tend to burn in air with extremely smoky flames - full of carbon particles.

You almost invariably get incomplete combustion, and the arenes can be recognized by the smokiness of their flames.

Hydrogenation

Hydrogenation is an addition reaction in which hydrogen atoms are added all the way around the benzene ring. A cycloalkane is formed. For example:

With benzene:

$$C_6H_6 + 3H_2 \longrightarrow C_6H_{12}$$

and methylbenzene:

$$C_6H_5CH_3 + 3H_2 \longrightarrow C_6H_{11}CH_3$$

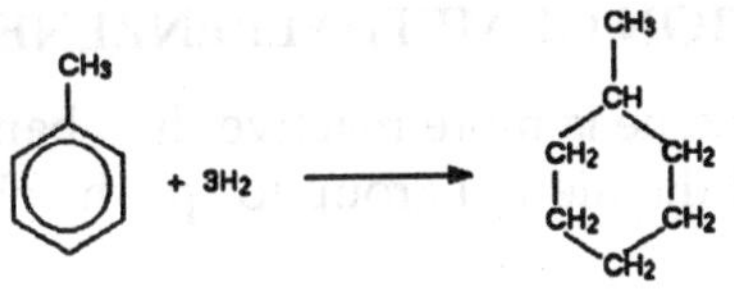

These reactions destroy the electron delocalization in the original benzene ring, because those electrons are being used to form bonds with the new hydrogen atoms.

Although the reactions are exothermic overall because of the strengths of all the new carbon-hydrogen bonds being made, there is a high activation barrier to the reaction.

The reactions are done using the same finely divided nickel catalyst that is used in hydrogenating alkenes and at similar temperatures (around 150°C), but the pressures used tend to be higher.

SULPHONATION

Sulphonation involves replacing one of the hydrogens on a benzene ring by the sulphonic acid group, $-SO_3H$.

SULPHONATION OF BENZENE

There are two equivalent ways of sulphonating benzene:

- Heat benzene under reflux with concentrated sulphuric acid for several hours.
- Warm benzene under reflux at 40°C with fuming sulphuric acid for 20 to 30 minutes. Fuming sulphuric

acid, $H_2S_2O_7$, can usefully be thought of as a solution of sulphur trioxide in concentrated sulphuric acid.

$$C_6H_6 + H_2SO_4 \longrightarrow C_6H_5SO_3H + H_2O$$

Or:

$$C_6H_6 + H_2SO_4 \longrightarrow C_6H_5SO_3H + H_2O$$

The product is benzenesulphonic acid.

SULPHONATION OF METHYLBENZENE

Methylbenzene is more reactive than benzene because of the tendency of the methyl group to "push" electrons towards the ring.

The effect of this greater reactivity is that methylbenzene will react with fuming sulphuric acid at 0°C, and with concentrated sulphuric acid if they are heated under reflux for about 5 minutes.

As well as the effect on the rate of reaction, with methylbenzene you also have to think about where the sulphonic acid group ends up on the ring relative to the methyl group.

Methyl groups have a tendency to "direct" new groups into the 2- and 4- positions on the ring (assuming the methyl group is in the 1- position). Methyl groups are said to be 2,4-*directing*. So you get a mixture which mainly consists of two isomers. Only about 5 - 10% of the 3- isomer is formed.

The main reactions are:

$$C_6H_5CH_3 + H_2SO_4 \longrightarrow 2\text{-}CH_3C_6H_4SO_3H + H_2O$$

and:

$$C_6H_5CH_3 + H_2SO_4 \longrightarrow 4\text{-}CH_3C_6H_4SO_3H + H_2O$$

In the case of sulphonation, the exact proportion of the isomers formed depends on the temperature of the reaction. As the temperature increases, you get increasing proportions of the 4- isomer and less of the 2- isomer.

This is because sulphonation is reversible. The sulphonic acid group can fall off the ring again, and reattach somewhere else. This tends to favour the formation of the most thermodynamically stable isomer. This interchange happens more at higher temperatures.

The 4- isomer is more stable because there is no cluttering in the molecule as there would be if the methyl group and sulphonic acid group were next door to each other.

Side Chain Oxidation in Alkylbenzenes

An alkylbenzene is simply a benzene ring with an alkyl group attached to it. Methylbenzene is the simplest alkylbenzene.

Alkyl groups are usually fairly resistant to oxidation. However, when they are attached to a benzene ring, they are easily oxidised by an alkaline solution of potassium manganate(VII) (potassium permanganate).

Methylbenzene is heated under reflux with a solution of potassium manganate(VII) made alkaline with sodium carbonate. The purple colour of the potassium manganate(VII) is eventually replaced by a dark brown precipitate of manganese(IV) oxide.

The mixture is finally acidified with dilute sulphuric acid.

Overall, the methylbenzene is oxidised to benzoic acid.

$$C_6H_5CH_3 \xrightarrow[\text{(b) dil } H_2SO_4]{\text{(a) alkaline } KMnO_4} C_6H_5COOH$$

Interestingly, *any* alkyl group is oxidised back to a -COOH group on the ring under these conditions. So, for example, propylbenzene is also oxidised to benzoic acid.

$$C_6H_5CH_2CH_2CH_3 \xrightarrow[\text{(a) dil } H_2SO_4]{\text{(a) alkaline } KMnO_4} C_6H_5COOH$$

NITRATION OF BENZENE AND METHYLBENZENE

Nitration of Benzene

Nitration happens when one (or more) of the hydrogen atoms on the benzene ring is replaced by a nitro group, NO_2.

Benzene is treated with a mixture of concentrated nitric acid and concentrated sulphuric acid at a temperature not exceeding 50°C. The mixture is held at this temperature for about half an hour. Yellow oily nitrobenzene is formed.

$$C_6H_6 + HNO_3 \longrightarrow C_6H_5NO_2 + H_2O$$

You could write this in a more condensed form as:

$$C_6H_6 + HNO_3 \longrightarrow C_6H_5NO_2 + H_2O$$

The concentrated sulphuric acid is acting as a catalyst and so isn't written into the equations.

At higher temperatures there is a greater chance of getting more than one nitro group substituted onto the ring. You will get a certain amount of 1,3-dinitrobenzene formed even at 50°C. Some of the nitrobenzene formed reacts with the nitrating mixture of concentrated acids.

$$C_6H_5NO_2 + HNO_3 \longrightarrow C_6H_4(NO_2)_2 + H_2O$$

Notice that the new nitro group goes into the 3 position on the ring. Nitro groups "direct" new groups into the 3 and 5 positions.

It is also possible to get a third nitro group attached to the ring (in the 5 position). However, nitro groups make the ring much less reactive than the original benzene ring. Two nitro groups on the ring make its reactions so slow that virtually no trinitrobenzene is produced under these conditions.

NITRATION OF METHYLBENZENE (TOLUENE)

Methylbenzene reacts rather faster than benzene - in nitration, the reaction is about 25 times faster. That means that you would use a lower temperature to prevent more than one nitro group being substituted - in this case, 30°C rather than 50°C. Apart from that, the reaction is just the same - using the same nitrating mixture of concentrated sulphuric and nitric acids.

You get a mixture of mainly two isomers formed: 2-nitromethylbenzene and 4-nitromethylbenzene. Only about 5% of the product is 3-nitromethylbenzene. Methyl groups are said to be *2,4-directing*.

For 2-nitromethylbenzene:

NO_2 + HNO_3 ⟶ NO_2, NO_2 + H_2O

and for 4-nitromethylbenzene:

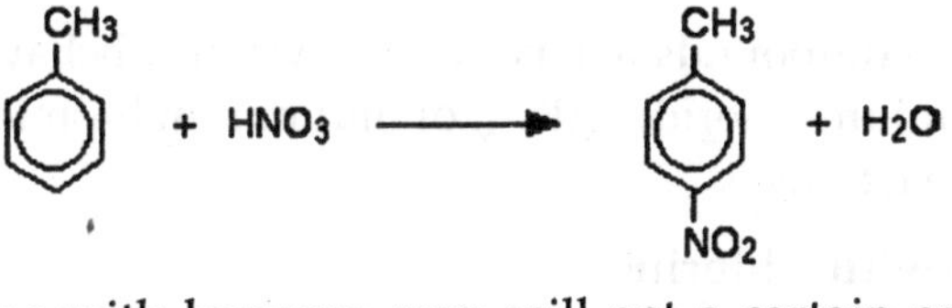

CH_3 + HNO_3 ⟶ CH_3, NO_2 + H_2O

Just as with benzene, you will get a certain amount of dinitro compound formed under the conditions of the reaction, but virtually no trinitro product because the reactivity of the ring decreases for every nitro group added. From an experimental point of view this is just as well. Trinitromethylbenzene used to be called trinitrotoluene or TNT!

The reactivity of a benzene ring is governed by the electron density around the ring. Methyl groups tend to "push" electrons towards the ring - increasing the density, and so making the ring more attractive to attacking reagents.

HALOGENATION OF BENZENE AND METHYLBENZENE

HALOGENATION OF BENZENE

Substitution Reactions

Benzene reacts with chlorine or bromine in the presence of a catalyst, replacing one of the hydrogen atoms on the ring by a chlorine or bromine atom.

The reactions happen at room temperature. The catalyst is either aluminium chloride (or aluminium bromide if you are reacting benzene with bromine) or iron.

Strictly speaking iron isn't a catalyst, because it gets permanently changed during the reaction. It reacts with some of the chlorine or bromine to form iron(III) chloride, $FeCl_3$, or iron(III) bromide, $FeBr_3$.

$$2Fe + 3Cl_2 \longrightarrow 2FeCl_3$$

$$2Fe + 3Br_2 \longrightarrow 2FeBr_3$$

These compounds act as the catalyst and behave exactly like aluminium chloride, $AlCl_3$, or aluminium bromide, $AlBr_3$, in these reactions.

Reaction with Chlorine

The reaction between benzene and chlorine in the presence of either aluminium chloride or iron gives chlorobenzene.

$$C_6H_6 \text{ (benzene ring)} + Cl_2 \longrightarrow C_6H_5Cl \text{ (ring with Cl)} + HCl$$

or, written more compactly:

$$C_6H_6 + Cl_2 \longrightarrow C_6H_5Cl + HCl$$

Reaction with Bromine

The reaction between benzene and bromine in the presence of either aluminium bromide or iron gives bromobenzene. Iron is usually used because it is cheaper and more readily available.

Br

+ Br_2 ⟶ + HBr

or

$$C_6H_6 + Br_2 \rightarrow C_6H_5Br + HBr$$

Addition Reactions

In the presence of ultraviolet light (but without a catalyst present), hot benzene will also undergo an addition reaction with chlorine or bromine. The ring delocalization is permanently broken and a chlorine or bromine atom adds on to each carbon atom.

For example, if you bubble chlorine gas through hot benzene exposed to UV light for an hour, you get 1,2,3,4,5,6-hexachlorocyclohexane.

+ $3Cl_2$ $\xrightarrow{\text{UV, heat}}$

Bromine would behave similarly.

The chlorines and hydrogens can stick up and down at random above and below the ring and this leads to a number of geometric isomers. Although there aren't any carbon-carbon double bonds, the bonds are still "locked" and unable to rotate.

One of these isomers was once commonly used as an insecticide known variously as BHC, HCH and Gammexane. This is one of the "chlorinated hydrocarbons" which caused so much environmental harm.

HALOGENATION OF METHYLBENZENE

Substitution Reactions

It is possible to get two quite different substitution reactions between methylbenzene and chlorine or bromine depending on the conditions used. The chlorine or bromine can substitute into the ring or into the methyl group.

Substitution into the Ring

Substitution in the ring happens in the presence of aluminium chloride (or aluminium bromide if you are using bromine) or iron, and in the absence of UV light. The reactions happen at room temperature.

This is exactly the same as the reaction with benzene, except that you have to worry about where the halogen atom attaches to the ring relative to the position of the methyl group.

Methyl groups are 2,4-directing, which means that incoming groups will tend to go into the 2 or 4 positions on the ring - assuming the methyl group is in the 1 position. In other words, the new group will attach to the ring next door to the methyl group or opposite it.

With chlorine, substitution into the ring gives a mixture of 2-chloromethylbenzene and 4-chloromethylbenzene.

CH_3 + Cl_2 ⟶ CH_3 Cl + HCl

CH_3 + Cl_2 ⟶ CH_3 Cl + HCl

With bromine, you would get the equivalent bromine compounds.

Substitution into the Methyl Group

If chlorine or bromine react with boiling methylbenzene in the absence of a catalyst but in the presence of UV light, substitution happens in the methyl group rather than the ring.

For example, with chlorine (bromine would be similar):

CH_3 + Cl_2 ⟶ CH_2Cl + HCl

The organic product is (chloromethyl)benzene. The brackets in the name emphasize that the chlorine is part of the attached methyl group, and isn't on the ring.

One of the hydrogen atoms in the methyl group has been replaced by a chlorine atom. However, the reaction doesn't stop there, and all three hydrogens in the methyl group can in turn be replaced by chlorine atoms.

That means that you could also get (dichloromethyl) benzene and (trichloromethyl) benzene as the other hydrogen atoms in the methyl group are replaced one at a time.

$C_6H_5CH_2Cl + Cl_2 \longrightarrow C_6H_5CHCl_2 + HCl$

$C_6H_5CHCl_2 + Cl_2 \longrightarrow C_6H_5CCl_3 + HCl$

If you use enough chlorine you will eventually get (trichloromethyl)benzene, but any other proportions will always lead to a mixture of products.

Addition Reactions

There nothing similar to the reaction between benzene and chlorine in which six chlorine atoms *add* around the ring.

This is because chlorine adds to benzene in the presence of ultraviolet light. With methylbenzene under those conditions, you get substitution in the methyl group. That is energetically easier because it doesn't involve breaking the delocalized electron system.

Once all the hydrogens in the methyl group had been substituted, *perhaps* you might then get addition to the ring as well.

FRIEDEL-CRAFTS REACTIONS OF BENZENE AND METHYLBENZENE

Friedel-Crafts Acylation of Benzene

The most commonly used acyl group is CH_3CO-. This is called the ethanoyl group, and in this case the reaction is sometimes called "ethanoylation". In the example which follows we are substituting a CH_3CO- group into the ring, but you could equally well use any other acyl group.

The most reactive substance containing an acyl group is an acyl chloride (also known as an acid chloride). These have the general formula RCOCl.

Benzene is treated with a mixture of ethanoyl chloride, CH_3COCl, and aluminium chloride as the catalyst. The mixture is heated to about 60°C for about 30 minutes.

A ketone called phenylethanone (old name: acetophenone) is formed.

$$C_6H_6 + CH_3C(=O)Cl \longrightarrow C_6H_5C(=O)CH_3 + HCl$$

or, if you want a more compact form:

$$C_6H_6 + CH_3COCl \longrightarrow C_6H_5COCH_3 + HCl$$

The aluminium chloride isn't written into these equations because it is acting as a catalyst. If you wanted to include it, you could write AlCl3 over the top of the arrow (see below).

Friedel-Crafts Acylation of Methylbenzene (Toluene)

The reaction is just the same with methylbenzene except that you have to worry about where the acyl group attaches to the ring relative to the methyl group.

Normally, the methyl group in methylbenzene directs new groups into the 2- and 4- positions (assuming the methyl group is in the 1- position). In acylation, though, virtually all the substitution happens in the 4- position.

$$C_6H_5CH_3 + CH_3COCl \xrightarrow{AlCl_3} CH_3C_6H_4COCH_3 + HCl$$

FRIEDEL-CRAFTS ALKYLATION OF BENZENE AND METHYLBENZENE

Friedel-Crafts Alkylation of Benzene

Alkylation means substituting an alkyl group into something - in this case into a benzene ring. A hydrogen on the ring is replaced by a group like methyl or ethyl and so on.

Benzene reacts at room temperature with a chloroalkane (for example, chloromethane or chloroethane) in the presence of aluminium chloride as a catalyst. On this page, we will look at substituting a methyl group, but any other alkyl group could be used in the same way.

Substituting a methyl group gives methylbenzene.

$$C_6H_6 + CH_3Cl \longrightarrow C_6H_5CH_3 + HCl$$

or:

$$C_6H_6 + CH_3Cl \longrightarrow C_6H_5CH_3 + HCl$$

Friedel-Crafts Alkylation of Methylbenzene (Toluene)

Again, the reaction is just the same with methylbenzene except that you have to worry about where the alkyl group attaches to the ring relative to the methyl group.

Unfortunately this is a problem! Where the incoming alkyl group ends up depends to a large extent on the temperature of the reaction.

At 0°C, substituting methyl groups into methylbenzene, you get a mixture of the 2-.3- and 4- isomers in the proportion 54%/17%/29%. That's a higher proportion of the 3- isomer than you might expect.

At 25°C, the proportions change to 3%/69%/28%. In other words the proportion of the 3- isomer has increased even more. Raise the temperature some more and the trend continues.

FRIEDEL-CRAFTS ALKYLATION INDUSTRIALLY

Manufacture of Ethylbenzene

Ethylbenzene is an important industrial chemical used to make styrene (phenylethene), which in turn is used to make polystyrene - poly(phenylethene).

It is manufactured from benzene and ethene. There are several ways of doing this, some of which use a variation on Friedel-Crafts alkylation. The reaction is done in the liquid state. Ethene is passed through a liquid mixture of benzene, aluminium chloride and a catalyst promoter which might be chloroethane or hydrogen chloride. We are going to assume it is HCl.

Promoters are used to make catalysts work better.

There are two variants on the process. One (the Union Carbide/Badger process) uses a temperature no higher than 130°C and a pressure just high enough to keep everything liquid.

The other (the Monsanto process) uses a slightly higher temperature of 160°C which needs less catalyst.

(benzene) + $CH_2{=}CH_2$ ⟶ (benzene ring with CH_2CH_3)

or

$$C_6H_6 + CH_2{=}CH_2 \longrightarrow C_6H_5CH_2CH_3$$

Again, the aluminium chloride and HCl aren't written into these equations because they are acting as catalysts. If you wanted to include them, you could write $AlCl_3$ and HCl over the top of the arrow.

Chapter 17

Aryl Halides (Halogenoarenes)

INTRODUCTION

Structure of chlorobenzene

We'll look in some detail at the structure of chlorobenzene. Bromobenzene and iodobenzene are just the same.

The simplest way to draw the structure of chlorobenzene is:

but to understand chlorobenzene properly, you need to go a bit deeper than this.

BONDING

There is an interaction between the delocalized electrons in the benzene ring and one of the lone pairs on the chlorine atom. This overlaps with the delocalized ring electron system. This delocalization is by no means complete, but it does have a significant effect on the properties of both the carbon-chlorine bond and the polarity of the molecule.

The delocalization introduces some extra bonding between the carbon and the chlorine, making the bond stronger. This has a major effect on the reactions of compounds like chlorobenzene.

There is also some movement of electrons away from the chlorine towards the ring. Chlorine is quite electronegative and usually draws electrons in the carbon-chlorine bond towards itself. In this case, this is offset to some extent by the movement of electrons back towards the ring in the delocalization. The molecule is less polar than you would otherwise have expected.

PREPARATION

Making Chlorobenzene

Benzene reacts with chlorine in the presence of a catalyst, replacing one of the hydrogen atoms on the ring by a chlorine atom.

The reaction happens at room temperature. The catalyst is either aluminium chloride or iron.

Strictly speaking iron isn't a catalyst, because it gets permanently changed during the reaction. It reacts with some of the chlorine to form iron(III) chloride, $FeCl_3$.

$$2Fe + 3Cl_2 \longrightarrow 2FeCl_3$$

This compound acts as the catalyst and behaves exactly like aluminium chloride, $AlCl_3$, in this reaction.

The reaction between benzene and chlorine in the presence of either aluminium chloride or iron gives chlorobenzene.

$$C_6H_6 + Cl_2 \longrightarrow C_6H_5Cl + HCl$$

or, written more compactly:

$$C_6H_6 + Cl_2 \longrightarrow C_6H_5Cl + HCl$$

Making Bromobenzene

The reaction between benzene and bromine in the presence of either aluminium bromide (rather than aluminium chloride) or iron gives bromobenzene. Iron is usually used because it is cheaper and more readily available. If you use iron, it is first converted into iron(III) bromide by the reaction between the iron and bromine.

$$C_6H_6 + Br_2 \longrightarrow C_6H_5Br + HBr$$

or

$$C_6H_6 + Br_2 \longrightarrow C_6H_5Br + HBr$$

Making Iodobenzene

Iodobenzene can be made from the reaction of benzene with iodine if they are heated under reflux in the presence of concentrated nitric acid, but it is normally made from benzenediazonium chloride solution. That's what we will concentrate on here.

If you add cold potassium iodide solution to ice-cold benzenediazonium chloride solution, nitrogen gas is given off, and you get oily droplets of iodobenzene formed.

There is a simple reaction between the diazonium ions present in the benzenediazonium chloride solution and the iodide ions from the potassium iodide solution.

$$C_6H_5\overset{+}{N}{\equiv}N + I^- \longrightarrow C_6H_5I + N_2$$

PHYSICAL PROPERTIES

Boiling Point

Chlorobenzene, bromobenzene and iodobenzene are all oily liquids. The boiling points increase as the halogen atom gets bigger.

	Boiling Point (°C)
C_6H_5Cl	132
C_6H_5Br	156
C_6H_5I	189

The main attractions between the molecules will be Van der Waals dispersion forces. These increase as the number of electrons in the molecule increases. This is the reason that the boiling points increase as the halogen atom gets bigger.

There will also be permanent dipole-dipole attractions involved in the chlorobenzene and bromobenzene, but very little in the iodobenzene. Iodine has much the same electronegativity as carbon.

These dipole-dipole attractions must be very unimportant relative to the dispersion forces because the most polar molecule (the chlorobenzene) has the lowest boiling point of the three.

Solubility in Water

The aryl halides are insoluble in water. They are denser than water and form a separate lower layer.

The molecules are quite large compared with a water molecule. In order for chlorobenzene to dissolve it would have to break lots of existing hydrogen bonds between water molecules. You also have to break the quite strong Van der Waals dispersion forces between chlorobenzene molecules. Both of these cost energy.

The only new forces between the chlorobenzene and the water would be Van der Waals dispersion forces. These aren't as strong as hydrogen bonds (or the original dispersion forces in the chlorobenzene), and so you wouldn't get much energy released when they form.

It simply isn't energetically profitable for chlorobenzene (and the others) to dissolve in water.

REACTIONS OF ARYL HALIDES (HALOGENOARENES)

Nucleophilic Substitution in the Halogenoalkanes

Here is a quick summary of the two ways that halogenoalkanes can react with hydroxide ions. We'll compare these with the aryl halides afterwards.

The two different ways in which these reactions can happen depends on what kind of halogenoalkane you are talking about.

Here is the mechanism for the reaction involving bromoethane - a primary halogenoalkane. A hydroxide ion attacks the slightly positive carbon atom and pushes off the bromine as a bromide ion.

$$CH_3-\overset{\delta+}{CH_2}-\overset{\delta-}{Br} \longrightarrow CH_3-CH_2-OH + :Br^-$$
$$:\bar{O}H$$

A tertiary halogenoalkane reacts differently.

The mechanism this time involves an initial ionization of the halogenoalkane:

$$CH_3-\underset{CH_3}{\overset{CH_3}{C}}-Br \overset{slow}{\rightleftharpoons} CH_3-\underset{CH_3}{\overset{CH_3}{C^+}} + :Br^-$$

followed by a very rapid attack by the hydroxide ion on the carbocation (carbonium ion) formed:

$$CH_3-\underset{CH_3}{\overset{CH_3}{C^+}} \;\; :\bar{O}H \xrightarrow{fast} CH_3-\underset{CH_3}{\overset{CH_3}{C}}-OH$$

Nucleophilic Substitution in the Aryl Halides

Simple aryl halides like chlorobenzene are very resistant to nucleophilic substitution. It is possible to replace the chlorine by -OH, but only under very severe industrial conditions - for example at 200°C and 200 atmospheres.

In the lab, these reactions don't happen. There are two reasons for this - depending on which of the above mechanisms you are talking about.

The Extra Strength of the Carbon-halogen Bond in Aryl Halides

The carbon-chlorine bond in chlorobenzene is stronger than you might expect. There is an interaction between one of the lone pairs on the chlorine atom and the delocalized ring electrons, and this strengthens the bond.

Both of the mechanisms above involve breaking the carbon-halogen bond at some stage. The more difficult it is to break, the slower the reaction will be.

Repulsion by the Ring Electrons

This will only apply if the hydroxide ion attacked the chlorobenzene by a mechanism like the first one described above. In that mechanism, the hydroxide ion attacks the slightly positive carbon that the halogen atom is attached to.

If the halogen atom is attached to a benzene ring, the incoming hydroxide ion is going to be faced with the delocalized ring electrons above and below that carbon atom. The negative hydroxide ion will simply be repelled.

Chapter 18

Phenol

INTRODUCTION

Phenol is the simplest member of a family of compounds in which an -OH group is attached directly to a benzene ring.

Structure of Phenol

The simplest way to draw the structure of phenol is:

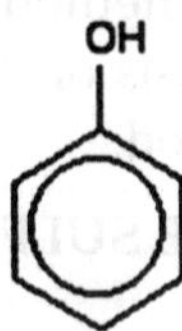

but to understand phenol properly, you need to go a bit deeper than this.

There is an interaction between the delocalized electrons in the benzene ring and one of the lone pairs on the oxygen atom. This has an important effect on both the properties of the ring and of the -OH group.

One of the lone pairs on the oxygen overlaps with the delocalized ring electron system. The donation of the oxygen's lone pair into the ring system increases the electron density around the ring. That makes the ring much more reactive than it is in benzene itself. It also helps to make the -OH group's hydrogen a lot more acidic than it is in alcohols. That will also be explored elsewhere in this section.

PREPARATION

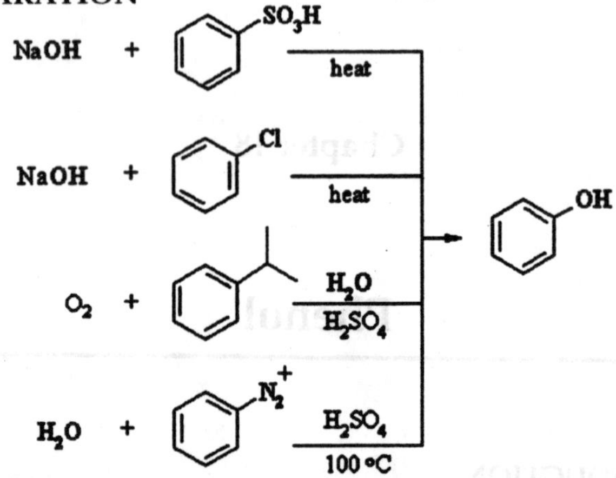

- Reaction of benzene sulfonic acid with base
- Reaction of chlorobenzene with base
- Acidic oxidation of cumene
- Hydrolysis of diazonium salts

Note: The first three methods are primarily industrial methods while the hydrolysis of diazonium salts is the most important laboratory method.

REACTION OF BENZENE SULFONIC ACID WITH HYDROXIDE

Reaction Type : Nucleophilic Aromatic Substitution

Summary

- This is a useful industrial method for the synthesis of phenol.
- Reagents : Usually NaOH, 300-350 °C followed by an acidic work-up to neutralize the phenoxide.
- This represents the oldest method for the industrial preparation of phenol.
- The reaction occurs via the addition-elimination mechanism with SO_3^{2-} functioning as the leaving group.

BASE HYDROLYSIS OF CHLOROBENZENE

Reaction Type : Nucleophilic Aromatic Substitution

Summary

- This is a useful industrial method for the synthesis of phenol.
- Reagents : Usually NaOH, 350 °C followed by an acidic work-up to neutralize the phenoxide.
- The reaction occurs via the elimination-addition mechanism via a benzyne intermediate.
- Elimination of HCl creates the benzyne that then undergoes addition of H_2O to produce the phenol.

ACIDIC OXIDATION OF CUMENE

Reaction type : Oxidation

Summary

- This industrial method is the primary source of phenol.
- Oxidation at the benzylic position gives the hydroperoxide which cleaves to give phenol and acetone.
- Cheap reagents and important products make the process attractive.

Preparation of Phenols from Aryl Diazonium Salts

Summary

- Aryl diazonium salts can be converted into phenols using H_2O/H_2SO_4/heat
- Aryl diazonium salts are prepared by reaction of aryl amines with nitrous acid, HNO_2

PHYSICAL PROPERTIES

Pure phenol is a white crystalline solid, smelling of disinfectant. It has to be handled with great care because it causes immediate white blistering to the skin. The crystals are often rather wet and discoloured.

Melting and Boiling Points

It is useful to compare phenol's melting and boiling points with those of methylbenzene (toluene). Both molecules contain the same number of electrons and are a very similar shape. That means that the intermolecular attractions due to Van der Waals dispersion forces are going to be very similar.

	Melting Point (°C)	*Boiling Point (°C)*
C_6H_5OH	40 - 43	182
C_6H_5CH3	-95.0	111

The reason for the higher values for phenol is in part due to permanent dipole-dipole attractions due to the electronegativity of the oxygen - but is mainly due to hydrogen bonding.

Hydrogen bonds can form between a lone pair on an oxygen on one molecule and the hydrogen on the -OH group of one of its neighbours.

Solubility in Water

Phenol is moderately soluble in water - about 8 g of phenol will dissolve in 100 g of water.

If you try to dissolve more than this, you get two layers of *liquid*. The top layer is a solution of phenol in water, and the bottom one a solution of water in phenol. The solubility behaviour of phenol and water is complicated.

Phenol is somewhat soluble in water because of its ability to form hydrogen bonds with the water.

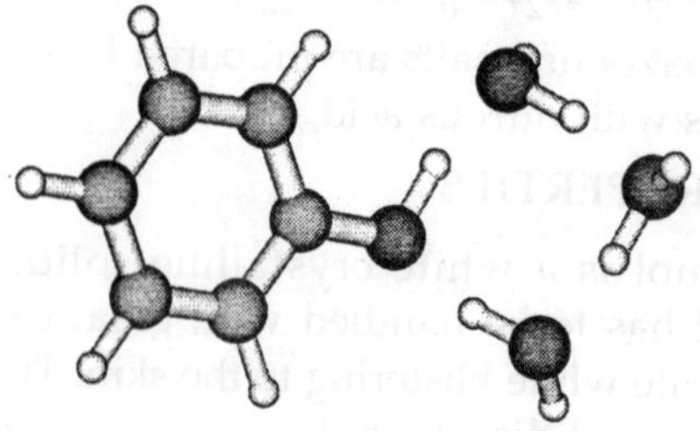

Fig. Hydrogen bonding in phenol

ACIDITY OF PHENOL

Unlike alcohols (which also contain an -OH group) phenol is a weak acid. A hydrogen ion can break away from the -OH group and transfer to a base.

For example, in solution in water:

$$C_6H_5OH + H_2O \rightleftharpoons C_6H_5O^- + H_3O^+$$

a phenoxide ion

Phenol is a very weak acid and the position of equilibrium lies well to the left.

Phenol can lose a hydrogen ion because the phenoxide ion formed is stabilized to some extent. The negative charge on the oxygen atom is delocalized around the ring. The more stable the ion is, the more likely it is to form.

One of the lone pairs on the oxygen atom overlaps with the delocalized electrons on the benzene ring. This overlap leads to a delocalization which extends from the ring out over the oxygen atom. As a result, the negative charge is no longer entirely localized on the oxygen, but is spread out around the whole ion.

Spreading the charge around makes the ion more stable than it would be if all the charge remained on the oxygen.

However , oxygen is the most electronegative element in the ion and the delocalized electrons will be drawn towards it. That means that there will still be a lot of charge around the oxygen which will tend to attract the hydrogen ion back again.

That's why phenol is only a very weak acid.

PROPERTIES OF PHENOL AS AN ACID

With Indicators

The pH of a typical dilute solution of phenol in water is likely to be around 5 - 6 (depending on its concentration). That means that a very dilute solution isn't really acidic enough to turn litmus paper fully red. Litmus paper is blue at pH 8 and red at pH 5. Anything in between is going to show as some shade of "neutral".

With Sodium Hydroxide Solution

Phenol reacts with sodium hydroxide solution to give a colourless solution containing sodium phenoxide.

$$C_6H_5OH + NaOH \longrightarrow C_6H_5O^- Na^+ + H_2O$$

sodium phenoxide

In this reaction, the hydrogen ion has been removed by the strongly basic hydroxide ion in the sodium hydroxide solution.

With Sodium Carbonate or Sodium Hydrogencarbonate

Phenol isn't acidic enough to react with either of these. Or, looked at another way, the carbonate and hydrogencarbonate ions aren't strong enough bases to take a hydrogen ion from the phenol.

Unlike the majority of acids, phenol *doesn't* give carbon dioxide when you mix it with one of these.

This lack of reaction is actually useful. You can recognize phenol because:

- It is fairly insoluble in water.
- It reacts with sodium hydroxide solution to give a colourless solution (and therefore must be acidic).
- It doesn't react with sodium carbonate or hydrogencarbonate solutions (and so must be only very weakly acidic).

With Metallic Sodium

Acids react with the more reactive metals to give hydrogen gas. Phenol is no exception - the only difference is the slow reaction because phenol is such a weak acid.

Phenol is warmed in a dry tube until it is molten, and a small piece of sodium added. There is some fizzing as hydrogen gas is given off. The mixture left in the tube will contain sodium phenoxide.

$$2\ C_6H_5OH + 2Na \longrightarrow 2\ C_6H_5O^-Na^+ + H_2$$

sodium phenoxide

REACTIONS OF PHENOL

Combustion of Phenol

Phenol burns in a plentiful supply of oxygen to give carbon dioxide and water.

$$C_6H_5OH + 7O_2 \longrightarrow 6CO_2 + 3H_2O$$

However, for compounds containing benzene rings, combustion is hardly ever complete, especially if they are burnt in air. The high proportion of carbon in phenol means that you need a very high proportion of oxygen to phenol to get complete combustion. Look at the equation.

As a general rule, the hydrogen in a molecule tends to get what oxygen is available first, leaving the carbon to form carbon itself, or carbon monoxide, if there isn't enough oxygen to go round.

Phenol tends to burn in air with an extremely smoky flame - full of carbon particles.

ESTERIFICATION OF PHENOL

You will probably remember that you can make esters from *alcohols* by reacting them with carboxylic acids. You might expect phenol to be similar.

However, unlike alcohols, phenol reacts so slowly with carboxylic acids that you normally react it with acyl chlorides (acid chlorides) or acid anhydrides instead.

Making Esters from Phenol using an Acyl Chloride

A typical acyl chloride is ethanoyl chloride, CH_3COCl.

Phenol reacts with ethanoyl chloride at room temperature, although the reaction isn't as fast as the one between ethanoyl chloride and an alcohol. Phenyl ethanoate is formed together with hydrogen chloride gas.

$$CH_3COCl + C_6H_5OH \longrightarrow CH_3COOC_6H_5 + HCl$$

Sometimes it is necessary to modify the phenol first to make the reaction faster.

For example, benzoyl chloride has the formula C_6H_5COCl. The -COCl group is attached directly to a benzene ring. It is much less reactive than simple acyl chlorides like ethanoyl chloride.

In order to get a reasonably quick reaction with benzoyl chloride, the phenol is first converted into sodium phenoxide by dissolving it in sodium hydroxide solution.

$$C_6H_5OH + NaOH \longrightarrow C_6H_5O^-Na^+ + H_2O$$

The phenoxide ion reacts more rapidly with benzoyl chloride than the original phenol does, but even so you have to shake it with benzoyl chloride for about 15 minutes. Solid phenyl benzoate is formed.

$$C_6H_5COCl + C_6H_5O^-Na^+ \longrightarrow C_6H_5COOC_6H_5 + NaCl$$

Making Esters from Phenol using an Acid Anhydride

A typical acid anhydride is ethanoic anhydride, $(CH_3CO)_2O$.

The reactions of acid anhydrides are slower than the corresponding reactions with acyl chlorides, and you usually need to warm the mixture.

Again, you can react the phenol with sodium hydroxide solution first, producing the more reactive phenoxide ion.

If you simply use phenol and ethanoic anhydride, phenyl ethanoate is formed together with ethanoic acid.

$$(CH_3CO)_2O + C_6H_5OH \longrightarrow CH_3COOC_6H_5 + CH_3COOH$$

This reaction isn't important itself, but a very similar reaction is involved in the manufacture of aspirin.

If the phenol is first converted into sodium phenoxide by adding sodium hydroxide solution, the reaction is faster. Phenyl ethanoate is again formed, but this time the other product is sodium ethanoate rather than ethanoic acid.

$$(CH_3CO)_2O + C_6H_5O^-Na^+ \longrightarrow CH_3COOC_6H_5 + CH_3COO^-Na^+$$

Reaction with Iron(III) Chloride Solution

Iron(III) chloride is sometimes known as ferric chloride.

Iron(III) ions form strongly coloured complexes with several organic compounds including phenol. The colour of the complexes vary from compound to compound. The reaction with iron(III) chloride solution can be used as a test for phenol.

If you add a crystal of phenol to iron(III) chloride solution, you get an intense violet-purple solution formed.

RING REACTIONS OF PHENOL

HOW DOES THE -OH GROUP MODIFY THE RING REACTIONS?

Activation of the Ring

The -OH group attached to the benzene ring in phenol has the effect of making the ring much more reactive than it would otherwise be.

For example, as you will find below, phenol will react with a solution of bromine in water (bromine water) in the cold and in the absence of any catalyst. It also reacts with dilute nitric acid, whereas benzene itself needs a nitrating mixture of concentrated nitric acid and concentrated sulphuric acid.

One of the lone pairs on the oxygen atom in the -OH group overlaps with the delocalized ring electron system.

The donation of the oxygen's lone pair into the ring system increases the electron density around the ring.

A benzene ring undergoes substitution reactions in which the ring electrons are attacked by positive ions or the slightly positive parts of molecules. In other words, it undergoes electrophilic substitution. If you increase the electron density around the ring, it becomes even more attractive to incoming electrophiles. That's what happens in phenol.

Directing Effect of the -OH Group

The -OH group has more activating effect on some positions around the ring than others. That means that incoming groups will go into some positions much faster than they will into others.

The net effect of this is that the -OH group has a 2,4-*directing effect*. That means that incoming groups will tend to go into the 2- position (next door to the -OH group) or the 4- position (opposite the -OH group). You will get hardly any of the 3- isomer formed - it is produced too slowly.

SPECIFIC EXAMPLES

Reaction with Bromine Water

If bromine water is added to a solution of phenol in water, the bromine water is decolourised and a white precipitate is formed which smells of antiseptic.

The precipitate is 2,4,6-tribromophenol. (The smell is similar to the antiseptic TCP: 2,4,6-trichlorophenol.)

$$C_6H_5OH + 3Br_2 \longrightarrow C_6H_2Br_3OH + 3HBr$$

2,4,6-tribromophenol

Notice the multiple substitution around the ring - into all the activated positions. (The 6- position is, of course, just the same as the 2- position. Both are next door to the -OH group.)

Reactions with Nitric Acid

The reactions with nitric acid are complicated because nitric acid is an oxidising agent, and phenol is very easily oxidised to give complex tarry products. What follows misses all that complication out, and just concentrates on the ring substitution which happens as well.

With Dilute Nitric Acid

Phenol reacts with dilute nitric acid at room temperature to give a mixture of 2-nitrophenol and 4-nitrophenol.

OH (phenol) + HNO_3 ⟶ OH, NO_2 (2-nitrophenol) + H_2O

2-nitrophenol

OH (phenol) + HNO_3 ⟶ OH, NO_2 (4-nitrophenol) + H_2O

4-nitrophenol

With Concentrated Nitric Acid

With concentrated nitric acid, more nitro groups substitute around the ring to give 2,4,6-trinitrophenol (old name: picric acid).

OH (phenol) + $3HNO_3$ ⟶ OH, NO_2, NO_2, NO_2 (2,4,6-trinitrophenol) + $3H_2O$

2,4,6-trinitrophenol

Chapter 19

Phenylamine or Aniline

INTRODUCTION

Phenylamine is a primary amine - a compound in which one of the hydrogen atoms in an ammonia molecule has been replaced by a hydrocarbon group.

However, in comparison with simple primary amines like methylamine, the properties of phenylamine are slightly different. This is because the lone pair on the nitrogen atom interacts with the delocalized electrons in the benzene ring.

Structure of Phenylamine

The simplest way to draw the structure of phenylamine is:

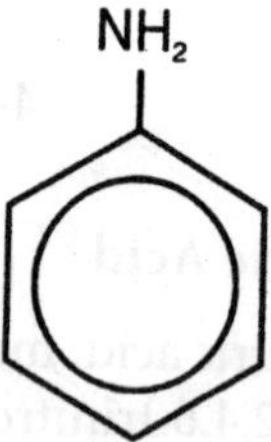

There is an interaction between the delocalized electrons in the benzene ring and the lone pair on the nitrogen atom. The lone pair overlaps with the delocalized ring electron system.

The donation of the nitrogen's lone pair into the ring system increases the electron density around the ring. That makes the ring much more reactive than it is in benzene itself.

It also reduces the availability of the lone pair on the nitrogen to take part in other reactions. In particular, it makes phenylamine much more weakly basic than primary amines where the -NH_2 group isn't attached to a benzene ring.

PREPARATION

FROM NITOBENZENE

Benzene to Nitrobenzene

Benzene is nitrated by replacing one of the hydrogen atoms on the benzene ring by a nitro group, NO_2.

The benzene is treated with a mixture of concentrated nitric acid and concentrated sulphuric acid at a temperature not exceeding 50°C. The mixture is held at this temperature for about half an hour. Yellow oily nitrobenzene is formed.

$$C_6H_6 + HNO_3 \longrightarrow C_6H_5NO_2 + H_2O$$

You could write this in a more condensed form as:

$$C_6H_6 + HNO_3 \longrightarrow C_6H_5NO_2 + H_2O$$

The concentrated sulphuric acid is acting as a catalyst and so isn't written into the equations.

The temperature is kept relatively low to prevent more than one nitro group being substituted onto the ring.

Nitrobenzene to Phenylamine

The conversion is done in two main stages:

Stage 1: Conversion of Nitrobenzene into Phenylammonium Ions

Nitrobenzene is reduced to phenylammonium ions using a mixture of tin and concentrated hydrochloric acid. The mixture is heated under reflux in a boiling water bath for about half an hour.

Under the acidic conditions, rather than getting phenylamine directly, you instead get phenylammonium ions formed. The lone pair on the nitrogen in the phenylamine picks up a hydrogen ion from the acid.

The electron-half-equation for this reaction is:

Notice the positively charged ion formed.

$$C_6H_5NO_2 + 7H^+ + 6e^- \longrightarrow C_6H_5NH_3^+ + 2H_2O$$

The nitrobenzene has been reduced by gaining electrons in the presence of the acid.

The electrons come from the tin, which forms both tin(II) and tin(IV) ions.

$$Sn \longrightarrow Sn^{2+} + 2e^-$$

$$Sn^{2+} \longrightarrow Sn^{4+} + 2e^-$$

Stage 2: conversion of the phenylammonium ions into phenylamine

All you need to do is to remove the hydrogen ion from the -NH3+ group.

Sodium hydroxide solution is added to the product of the first stage of the reaction.

$$C_6H_5NH_3^+ + OH^- \longrightarrow C_6H_5NH_2 + H_2O$$

The phenylamine is formed together with a complicated mixture of tin compounds from reactions between the sodium hydroxide solution and the complex tin ions formed during the first stage.

The phenylamine is finally separated from this mixture. The separation is long, tedious and potentially dangerous - involving steam distillation, solvent extraction and a final distillation.

PHYSICAL PROPERTIES

Pure phenylamine is a colourless liquid, but it darkens rapidly on exposure to light and air. It is normally a brown oily liquid.

Melting and Boiling Points

It is useful to compare phenylamine's melting and boiling points with those of methylbenzene (toluene). Both molecules contain a similar number of electrons and are a very similar shape. That means that the intermolecular attractions due to Van der Waals dispersion forces are going to be very similar.

	Melting Point (°C)	*Boiling Point (°C)*
$C_6H_5NH_2$	-6.2	184
$C_6H_5CH_3$	-95.0	111

The reason for the higher values for phenylamine is in part due to permanent dipole-dipole attractions due to the electronegativity of the nitrogen - but is mainly due to hydrogen bonding.

Hydrogen bonds can form between a lone pair on a nitrogen on one molecule and the hydrogen on the -NH2 group of one of its neighbours.

Solubility in Water

Phenylamine is slightly soluble in water - about 3.6 g (depending on where you get the data from!) of phenylamine will dissolve in 100 g of water at 20°C. Mixtures containing more phenylamine than this separate into two layers, with the phenylamine forming the bottom one.

Phenylamine is somewhat soluble in water because of its ability to form hydrogen bonds with the water.

However, the benzene rings in the phenylamine break more hydrogen bonds between water molecules than are reformed between water and the -NH2 groups. The water molecules also disrupt fairly strong Van der Waals attractions between the phenylamine molecules.

Both of these effects mean that dissolving phenylamine in water isn't very energetically profitable, and so stop the phenylamine from being very soluble.

PHENYLAMINE AS A BASE

Amines are bases because the lone pair of electrons on the nitrogen atom can accept a hydrogen ion - in other words, for exactly the same reason that ammonia is a base.

With phenylamine, the only difference is that it is a much weaker base than ammonia or an amine like ethylamine - for reasons that we will explore later.

The Reaction of Phenylamine with Acids

Phenylamine reacts with acids like hydrochloric acid in exactly the same way as any other amine. Despite the fact that the phenylamine is only a very weak base, with a strong acid like hydrochloric acid the reaction is completely straightforward.

Phenylamine is only very slightly soluble in water, but dissolves freely in dilute hydrochloric acid. A solution of a salt is formed - phenylammonium chloride.

If you just want to show the formation of the salt, you could write:

$$C_6H_5NH_2 + HCl \longrightarrow C_6H_5NH_3^+Cl^-$$

or if you want to emphasize the fact that the phenylamine is acting as a base, you could most simply use:

$$C_6H_5NH_2(I) + H^+(aq) \longrightarrow C_6H_5NH_3^+(aq)$$

Getting the Phenylamine Back from its Salt

To get the phenylamine back from the phenylammonium ion present in the salt, all you have to do is to take the hydrogen ion away again. You can do that by adding any stronger base.

Normally, you would choose sodium hydroxide solution.

$$C_6H_5NH_3(aq) + OH^+(aq) \longrightarrow C_6H_5NH_2(I)+H_2O(I)$$

The phenylamine is formed first as an off-white emulsion - tiny droplets of phenylamine scattered throughout the water. This then settles out to give an oily bottom layer of phenylamine under the aqueous layer.

Reaction of Phenylamine with Water

This is where it is possible to tell that phenylamine is a much weaker base than ammonia and the aliphatic amines like methylamine and ethylamine.

Phenylamine reacts reversibly with water to give phenylammonium ions and hydroxide ions.

$$C_6H_5NH_{2(aq)} + H_2O(l) \rightleftharpoons C_6H_5NH_3^+{}_{(aq)} + OH^-{}_{(aq)}$$

The position of equilibrium lies well to the left of the corresponding ammonia or aliphatic amine equilibria - which means that not many hydroxide ions are formed in the solution.

The effect of this is that the pH of a solution of phenylamine will be quite a bit lower than a solution of ammonia or one of the aliphatic amines of the same concentration. For example, a 0.1 M phenylamine solution has a pH of about 9 compared to a pH of about 11 for 0.1 M ammonia solution.

Why is Phenylamine such a Weak Base?

Amines are bases because they pick up hydrogen ions on the lone pair on the nitrogen atom. In phenylamine, the attractiveness of the lone pair is lessened because of the way it interacts with the ring electrons.

The lone pair on the nitrogen touches the delocalized ring electrons and becomes delocalized with them.

That means that the lone pair is no longer fully available to combine with hydrogen ions. The nitrogen is still the most electronegative atom in the molecule, and so the delocalized electrons will be attracted towards it, but the electron density around the nitrogen is nothing like it is in, say, an ammonia molecule.

The other problem is that if the lone pair is used to join to a hydrogen ion, it is no longer available to contribute to the delocalization. That means that the delocalization would have to be disrupted if the phenylamine acts as a base. Delocalization makes molecules more stable, and so disrupting the delocalization costs energy and won't happen easily.

Taken together - the lack of intense charge around the nitrogen, and the need to break some delocalization - means that phenylamine is a very weak base indeed.

REACTIONS OF PHENYLAMINE

ACYLATION

Reactions with Acyl Chlorides and with Acid Anhydrides

We'll take ethanoyl chloride as a typical acyl chloride, and ethanoic anhydride as a typical acid anhydride. The important product of the reaction of phenylamine with either of these is the same.

Phenylamine reacts vigorously in the cold with ethanoyl chloride to give a mixture of solid products - ideally white, but usually stained brownish. A mixture of N-phenylethanamide (old name: acetanilide) and phenylammonium chloride is formed.

The overall equation for the reaction is:

$$CH_3COCl + 2C_6H_5NH_2 \longrightarrow CH_3CONHC_6H_5 + C_6H_5NH_3^+Cl^-$$

With ethanoic anhydride, heat is needed. In this case, the products are a mixture of N-phenylethanamide and phenylammonium ethanoate.

$$(CH_3CO)_2O + 2C_6H_5NH_2 \longrightarrow CH_3CONHC_6H_5 + CH_3COO^{-+}NH_3C_6H_5$$

The main product molecule (the N-phenylethanamide) is often drawn looking like this:

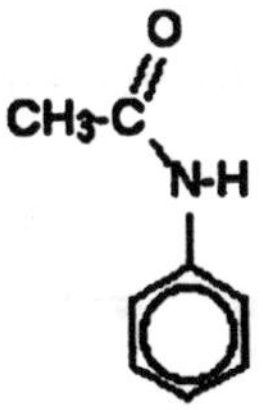

$$CH_3-C(=O)-NH-C_6H_5$$

If you stop and think about it, this is obviously the same molecule as in the equation above, but it stresses the phenylamine part of it much more.

Looking at it this way, notice that one of the hydrogens of the $-NH_2$ group has been replaced by an acyl group - an alkyl group attached to a carbon-oxygen double bond.

You can say that the phenylamine has been *acylated* or has undergone *acylation.*

Because of the nature of this particular acyl group, it is also described as *ethanoylation.* The hydrogen is being replaced by an ethanoyl group, CH_3CO-.

Reaction with Halogenoalkanes

This is another reaction of phenylamine as a nucleophile, and again there is no essential difference between its reactions and those of aliphatic amines.

Taking bromoethane as a typical halogenoalkane, the reaction with phenylamine happens in the same series of complicated steps as with any other amine.

We'll just look at the first step.

On heating, the bromoethane and phenylamine react to give a mixture of a salt of a secondary amine and some free secondary amine. In this case, you would first get N-ethylphenylammonium bromide:

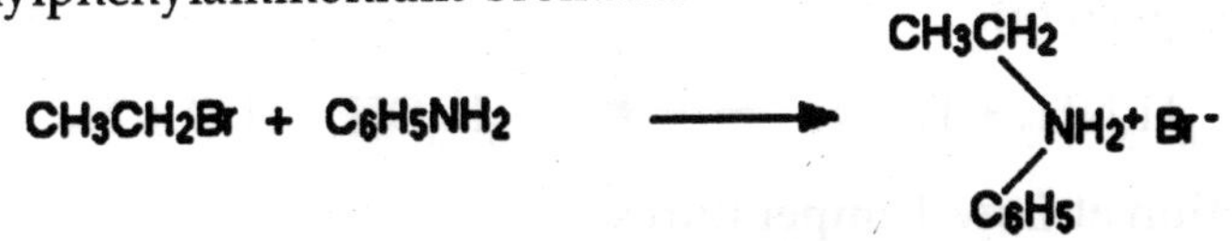

$$CH_3CH_2Br + C_6H_5NH_2 \longrightarrow (CH_3CH_2)(C_6H_5)NH_2^+ \; Br^-$$

but this would instantly be followed by a reversible reaction in which some unreacted phenylamine would take a hydrogen ion from the salt to give some free secondary amine: N-ethylphenylamine.

$$(CH_3CH_2)(C_6H_5)NH_2^+ Br^- + C_6H_5NH_2 \rightleftharpoons (CH_3CH_2)(C_6H_5)NH + C_6H_5NH_3^+ Br^-$$

The reaction wouldn't stop there. You will get further reactions to produce a tertiary amine and its salt, and eventually a quaternary ammonium compound.

MAKING DIAZONIUM SALTS FROM PHENYLAMINE

Reactions of Phenylamine with Nitrous Acid

Nitrous acid (also known as nitric(III) acid) has the formula HNO_2. It is sometimes written as HONO to show the way it is joined up.

Nitrous acid decomposes very readily and is always made *in situ*. In the case of its reaction with phenylamine, the phenylamine is first dissolved in hydrochloric acid, and then a solution of sodium or potassium nitrite is added. The reaction between the hydrochloric acid and the nitrite ions produces the nitrous acid.

You get the reaction:

$$H^+_{(aq)} + NO_2^-(aq) \rightleftharpoons HNO_{2(aq)}$$

Because nitrous acid is a weak acid, the position of equilibrium lies well the right.

Phenylamine reacts with nitrous acid differently depending on the temperature.

Reaction on Warming

If the mixture is warmed, you get a black oily product which contains phenol (amongst other things), and nitrogen gas is given off.

$$C_6H_5NH_2 + HNO_2 \longrightarrow C_6H_5OH + H_2O + N_2$$

Reaction at Low Temperatures

The solution of phenylamine in hydrochloric acid (phenylammonium chloride solution) is stood in a beaker of ice. The sodium or potassium nitrite solution is also cooled in the ice.

The solution of the nitrite is then added very slowly to the phenylammonium chloride solution - so that the temperature never goes above 5°C.

You end up with a solution containing benzenediazonium chloride:

Notice the positive charge on the nitrogen atom attached to the ring.

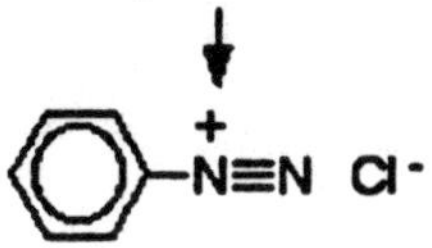

benzenediazonium chlooride

The positive ion, containing the -N2+ group, is known as a ***diazonium ion***. The "azo" bit of the name refers to nitrogen.

The ionic equation for the reaction is:

$$C_6H_5NH_2 + HNO_2 + H^+ \longrightarrow C_6H_5\text{-}\overset{+}{N}\equiv N + 2H_2O$$

Notice that the chloride ions from the acid aren't involved in this in any way. If you use hydrochloric acid, the solution will contain benzenediazonium chloride. If you used a different acid, you would just get a different salt - a sulphate or hydrogensulphate, for example, if you used sulphuric acid.

The reactions of a diazonium salt are always done with a freshly prepared solution made in this way. The solutions don't keep. Diazonium salts are very unstable and tend to be explosive as solids.